Progress
12 in Molecular and
Subcellular Biology

Edited by
P. Jeanteur, Y. Kuchino
W.E.G. Müller (Managing Editor)
P.L. Paine

With 20 Figures

Springer-Verlag
Berlin Heidelberg New York
London Paris Tokyo
Hong Kong Barcelona
Budapest

Prof. Dr. WERNER E.G. MÜLLER
Physiologisch-Chemisches Institut
Abt. Angewandte Molekularbiologie
Duesbergweg 6
6500 Mainz, FRG

Dr. YOSHIYUKI KUCHINO
National Cancer Center
Research Institute
Tsukiji 5-chome
Chuo-ku, Tokyo 104, Japan

Prof. Dr. PHILIPPE JEANTEUR
UA CNRS 1191
Laboratoire de Biologie Moléculaire
Université des Sciences et Technique
du Languedoc
34060 Montpellier Cedex, France

Prof. Dr. PHILIP L. PAINE
Department of
Biological Sciences
St. John's University
Grand Central & Utopia Parkways
Jamaica, New York 11439, USA

ISBN 3–540–53900–X Springer-Verlag Berlin Heidelberg New York
ISBN 0–387–53900–X Springer-Verlag New York Berlin Heidelberg

The Library of Congress has catalogued this serial publication as follows:
Library of Congress Catalog Card Number 75–79748

Typesetting: International Typesetters Inc., Makati, Metro Manila, Philippines
31/3145–543210 — Printed on acid-free paper

Contents

Synthesis of Small Nuclear RNAs
R. REDDY and R. SINGH

1	Introduction	1
2	Two Classes of U snRNA Genes	1
3	RNA Polymerases Transcribing U snRNA Genes	3
3.1	TMG-Capped snRNAs	3
3.2	MepppG-Capped snRNAs	4
4	Organization of U snRNA Genes	5
4.1	Human	5
4.2	Rodent	10
4.3	Chicken	10
4.4	*Xenopus*	10
4.5	*Drosophila*	11
4.6	Sea Urchin	11
4.7	*Trypanosome*	11
4.8	*C. elegans*	12
4.9	Yeast	12
4.10	Plants	12
4.11	Viral U RNAs	12
5	*Cis*-Acting Elements in U snRNA Genes	13
5.1	Initiation Nucleotide	13
5.2	Proximal Sequence Element (PSE)	13
5.3	TATA Box	14
5.4	Distal Sequence Element (DSE)	15
5.5	3'-End Formation	16
6	Interconversion of U snRNA Promoters	17
6.1	SnRNA and mRNA Gene Promoters Are Distinct	18
6.2	Transcription of snRNA Genes in Vitro and in Heterologous Systems	19
7	Regulation of snRNA Synthesis	19
8	Formation of Cap Structure in U snRNAs	20
8.1	Signal for U6 snRNA Capping	20
8.2	Post-Transcriptional Capping of U6 snRNA	22
8.3	Other snRNAs with mepppG Cap Structure	23

8.4   Functions of Cap Structures 24
9   Summary 25
References 26

The DNA-Activated Protein Kinase, DNA-PK
T.H. CARTER and C.W. ANDERSON

1   Nuclear Protein Kinases 37
2   Nucleic Acid Effects on Protein Kinase Activity 38
3   Detection of DNA-Stimulated Protein Phosphorylation
       in Cell Extracts 39
4   Purification of DNA-PK from HeLa Cells 42
5   Physical Characteristics of DNA-PK 43
6   Subcellular Localization 45
7   Phosphate Donor and Cofactor Requirements 45
8   Effects of Inhibitors 46
9   Substrate Specificity 47
9.1   Autophosphorylation 47
9.2   Substrate Preference and Phosphorylation Sites 48
10   Effects of Polynucleotides 49
11   Occurrence of DNA-PK in Other Cells 51
12   Comparison with Other Nuclear Protein Kinases 51
13   Conclusions and Future Directions 52
References 55

The Cytoskeleton During Early Development:
Structural Transformation and Reorganization of RNA and Protein
D.G. CAPCO and C.A. LARABELL

1   Introduction 59
2   The Cytoskeleton in the Early Development of Chordates 63
2.1   Ascidians 63
2.2   Amphibians 66
2.3   Mammals 71
3   The Cytoskeleton in the Early Development of Nonchordates 76
3.1   Annelids 76
3.2   Oligochaetes 77
3.3   Nematodes 78
3.4   Insects 78
3.5   Echinoderms 79
4   Conclusion 80
References 82

Developmental Regulations of Heat-Shock Protein Synthesis in Unstressed and Stressed Cells
O. Bensaude, V. Mezger, and M. Morange

1	Introduction	89
2	Expression of Heat-Shock Genes During Gametogenesis and Early Development in the Absence of Stress	90
2.1	An Ancient Developmental Response: Heat-Shock Protein Hyperexpression During Sporulation and Gametogenesis	90
2.2	Heat-Shock Proteins, First Major Products of the Zygotic Genome Transcription in Mammals	93
2.3	Constitutive Heat-Shock Protein Expression During Early Mouse Embryogenesis	94
2.4	Constitutive Heat-Shock Protein Expression in Mouse Embryonal Carcinoma (EC) Cells	94
2.5	High Levels of B2 Transcripts in Heat-Shocked Fibroblasts and in Undifferentiated Mouse Embryonic Cells	96
3	Heat-Shock Protein Synthesis in Differentiation Processes	96
3.1	Specificities of the Heat-Shock Protein Synthesis Associated with Blood Cell Differentiation	96
3.2	Hormone-Induced Heat-Shock Protein Expression	97
3.3	Entering or Leaving a Quiescent State	98
4	Deficient Heat-Shock Responses	100
4.1	Sporulation and Gametogenesis	100
4.2	Early Embryogenesis	100
4.3	Cultured Cells	101
5	Concluding Remarks	102
References		103

The Interactions of Water and Proteins in Cellular Function
J.G. Watterson

1	Introduction	113
2	The Cluster Model of Liquid Structure	116
2.1	Cluster Size	116
2.2	Cluster Energetics	119
3	The Domain Model of Protein Structure	123
3.1	Domain Size	123
3.2	Domain Energetics	127
References		131

Subject Index 135

Contributors

You will find the addresses at the beginning of the respective contribution.

Anderson, C.W. 37
Bensaude, O. 89
Capco, D.G. 59
Carter, T.H. 37
Larabell, C.A. 59
Mezger, V. 89
Morange, M. 89
Reddy, R. 1
Singh, R. 1
Watterson, J.G. 113

Synthesis of Small Nuclear RNAs

R. REDDY and R. SINGH[1]

1 Introduction

There are seven abundant and several less abundant capped small nuclear RNAs characterized in mammalian cells. These RNAs are all capped on their 5′ ends and were designated U snRNAs because the U1-U3 snRNAs initially studied were rich in uridylic acid (Hodnett and Busch 1968). These capped snRNAs play important roles in the processing of nuclear precursor mRNAs and precursor rRNAs (reviewed in Busch et al. 1982; Brunel et al. 1985; Green 1986; Padgett et al. 1986; Maniatis and Reed 1987; Guthrie and Patterson 1988; Steitz 1988; Steitz et al. 1988; Zieve and Sauterer 1990). The functions of the U snRNAs are summarized in Table 1. While the roles of U snRNAs in the processing of eukaryotic precursor RNAs are well established, U5 snRNA was recently shown to have the potential to transform cells in vitro (Hamada et al. 1989), suggesting multiple roles for the U snRNAs. Each HeLa cell contains a total of approximately 2–3 million copies of U snRNAs (Weinberg and Penman 1968), and it is estimated that each of the U1 and U2 snRNA genes is transcribed every 2–4 s, generating the large amounts of U snRNAs found in mammalian cells (Skuzeski et al. 1984; Mangin et al. 1986; reviewed in Dahlberg and Lund 1988); hence, the snRNA genes have very strong promoters compared to many other cellular genes.

2 Two Classes of U snRNA Genes

Based on the type of cap structure present on their 5′ ends, U snRNAs are divided into two classes. The trimethylguanosine (TMG) cap-containing U snRNAs include U1-U5 and U7-U14 snRNAs; and the methyl (mepppG) cap-containing U snRNAs include U6 and 7SK RNAs. These two cap structures are shown in Fig. 1. Although many small nuclear RNAs are capped, not all small RNAs in the nucleus contain cap structures. For instance, human RNaseP (H1) RNA (Baer et al. 1990) and 7SM (7–2/MRP) RNA (Hashimoto and Steitz 1983; Yuan et al. 1989) do not contain cap structures.

The first U snRNA gene to be isolated and characterized was the U3 snRNA gene from slime mold *Dictyostelium* (Wise and Weiner 1980). To date, approxi-

---

[1]Department of Pharmacology, Baylor College of Medicine, Houston, TX 77030, USA

Table 1. Functions of U snRNAs

RNA	Function	Reference
U1	Splicing of nuclear pre-mRNAs; binds specifically to the 5′ splice site	Mount et al. (1981); Steitz et al. (1988); Zhuang and Weiner (1986); Guthrie and Patterson (1988)
U2	Splicing of nuclear pre-mRNAs; binds specifically to the branch point	Parker et al. (1987); Steitz et al. (1988)
U3	Processing of pre-ribosomal RNA; binds specifically near the 5′ end of 45S RNA	Kass et al. (1990); Maser and Calvet (1989); Stroke and Weiner (1989)
U4	Splicing of nuclear pre-mRNAs; binds specifically to U6 snRNA	Berget and Robberson (1986); Black and Steitz (1986); Steitz et al. (1988)
U5	Splicing of nuclear pre-mRNAs; binds specifically to the 3′ splice site	Chabot et al. (1985); Tazi et al. (1986; Steitz et al. (1988)
U6	Splicing of nuclear pre-mRNAs; binds specifically to U4 snRNA	Berget and Robberson (1986); Black and Steitz (1986); Brown and Guthrie (1988); Steitz et al. (1988)
U7	3′-end formation of histone pre-mRNAs	Schaufele et al. (1986); Mowry and Steitz (1987)
U8	Not known; probably involved in pre-rRNA processing	Reddy et al. (1985); Tyc and Steitz (1989)
U11	Not known; probably involved in polyadenylation	Christofori and Keller (1988); Montzka and Steitz (1988)
U12	Not known; probably involved in nuclear pre-mRNA processing	Montzka and Steitz (1988)
U13	Not known; probably involved in pre-rRNA processing	Tyc and Steitz (1989)
U14	Disrupts production of 18S rRNA; probably involved in pre-rRNA processing	Li et al. (1990)
7SK	Not known; probably involved in nuclear pre-mRNA processing	Gupta et al. (1990b)

mately 100 snRNA genes from diverse species have been isolated and characterized. These include genes from human, rat, mouse, chicken, frog, *Drosophila*, sea urchin, *Trypanosome, C. elegans,* slime mold, yeast, and several plants. The U snRNA genes encoding the TMG-capped U1-U5 and U7-U14 snRNAs have many common features, as summarized in Table 2. U6 and 7SK RNA genes belong to another class, and they differ from the other U snRNA genes in several important respects; the main difference is that these two genes are transcribed by RNA polymerase III (pol III) in contrast to the transcription of TMG-capped snRNAs by RNA polymerase II (pol II).

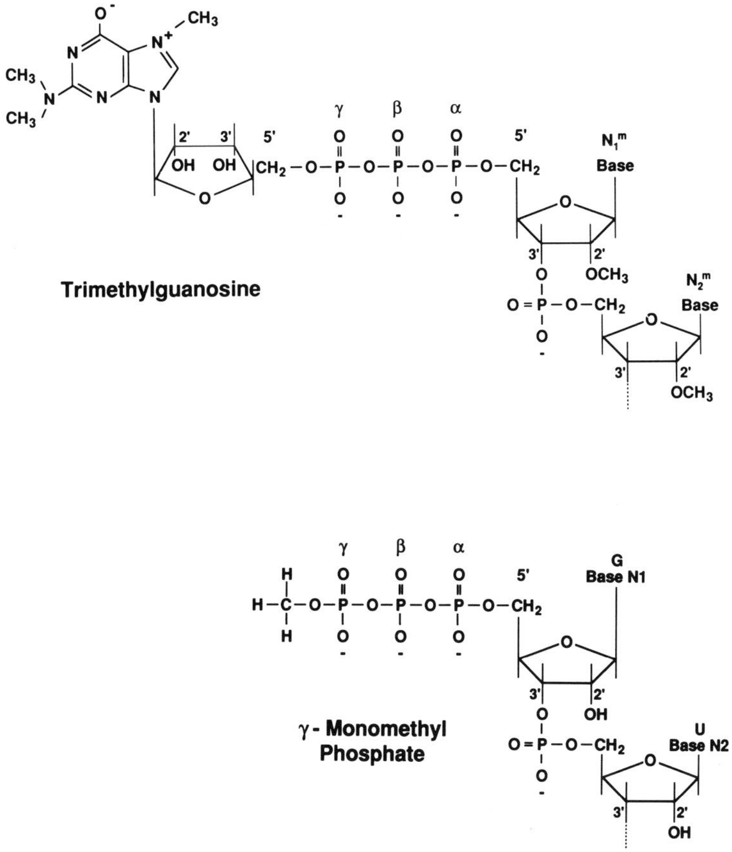

Fig. 1. Cap structures in U snRNAs. *Top*: TMG-cap structure found in U1–U5 and U7–U14 snRNAs. The 2′-O-methylations occur only in higher eukaryotes, such as rat and HeLa cells, but not in amoeba or dinoflagellates. *Bottom*: MepppG-cap structure found in U6 and 7SK snRNAs. Diagrammatic representation of the methylated γ-phosphate of the 5′ nucleotide G (nucleotide N1) of human U6 and 7SK snRNA. The 2′,3′, and 5′ represent the carbon moieties of the ribose sugar

3 RNA Polymerases Transcribing U snRNA Genes

3.1 TMG-Capped snRNAs

Several lines of evidence suggest that pol II is responsible for the synthesis of TMG-capped U snRNAs. (1) The synthesis of these RNAs is inhibited by low concentrations of α-amanitin in whole animals (Ro-Choi et al. 1976), cultured cells (Frederiksen et al. 1978; Chandrasekharappa et al. 1983), isolated nuclei (Roop et al. 1981; Lobo and Marzluff 1987), cell-free extracts (Morris et al. 1986; Lund and Dahlberg 1989; Southgate and Busslinger 1989), and frog oocytes (Murphy et al. 1982; Mattaj and Zeller 1983; Skuzeski et al. 1984; Reddy et al. 1987). (2) Gram-Jensen et al. (1979) used a cell line containing an altered pol II which is 800 times

Table 2. Two types of U snRNA genes

Characteristic	TMG-capped	mepppG-capped
Examples	U1-U5, U7-U14	U6, 7SK
5′-end cap	TMGpppA/G	CH_3-O-pppG/A
Synthesized by	Polymerase II (B)	Polymerase III (C)
Transcription factors	Share with mRNA genes	Share with mRNA genes
3′-end formation requires	3′ Box and a compatible snRNA promoter	T-stretch
Introns	Not present (with one exception)	Not present (with one exception)
Initiation nucleotide	Purine	Purine
PSE at –45 to –70[a]	Present and required	Present and required
DSE around –250[a]	Functions as enhancer	Functions as enhancer
Intragenic promoter	None	None

[a]PSE and DSE stand for Proximal and Distal Sequence Element, respectively. Mammalian U snRNA genes have sequences downstream of –50 that are important for PSE function (Murphy et al. 1987a; reviewed in Dahlberg and Lund 1988); U6 and 7SK snRNA genes, as well as plant U snRNA genes, contain an essential TATA-motif at –30 region (reviewed in Geiduschek and Tocchini-Valentini 1988; also see Sect. 5.3).

more resistant towards inhibition by α-amanitin than the wild-type enzyme. In these cells, the synthesis of U1, U2, and U3 snRNAs was not inhibited by high concentrations of α-amanitin. Furthermore, the synthesis of these U snRNAs is inhibited at nonpermissive temperature in the cells that contain a temperature-sensitive pol II (Hellung-Larsen et al. 1980). (3) The synthesis of U1 and U2 snRNAs is sensitive to 5,6-dichloro-1-β-D-ribofuranosyl benzimidazole, which is a specific inhibitor of transcription by pol II at low concentrations (Hellung-Larsen et al. 1981) and the primary transcripts of U1 snRNA, like mRNAs, are co-transcriptionally capped with m^7G (Eliceiri 1980; Skuzeski et al. 1984; Mattaj 1986). (4) Antibodies against the large subunit of the pol II inhibit the synthesis of U1 snRNA in the frog oocytes (Thompson et al. 1989). (5) Finally, Pol III is unlikely to be involved in the synthesis of TMG-capped U snRNAs because a U cluster (AUUUUG as Sm antigen-binding site) is present within the transcribed portion of a large number of these genes and this signal results in termination of pol III-mediated transcription. All these data show that TMG-capped snRNAs are synthesized by pol II or by an RNA polymerase closely related to pol II. Although studies have been carried out on the synthesis of only some TMG-capped snRNAs, it is likely that other TMG-capped snRNAs are also synthesized by pol II.

3.2 MepppG-Capped snRNAs

There is much evidence to support the involvement of pol III in the synthesis of U6 and 7SK RNAs. (1) Low concentrations of α-amanitin, sufficient to inhibit the synthesis of mRNAs and TMG-capped U snRNAs, had no inhibitory effect on the synthesis of U6 RNA in frog oocytes, or in vitro (Kunkel et al. 1986; Reddy et al. 1987; Krol et al. 1987), or in isolated nuclei (Kunkel et al. 1986). (2) In U6 snRNA genes, the signal for transcription termination is a T-cluster (Das et al. 1988) similar to the functional termination signal in 5S RNA gene (Bogenhagen and Brown 1981). (3) The transcription of U6 snRNA is competed by other pol III genes like 5S and tRNA genes both in vitro (Reddy et al. 1987) and in vivo (Carbon et al. 1987). (4) The U6 snRNA associates with La antigen (Rinke and Steitz 1985; Reddy et al. 1987) which may be a pol III transcription termination factor (Gottlieb and Steitz 1989). (5) A mutant yeast strain with temperature-sensitive defect in the large subunit of pol III, which results in defective transcription of tRNA and 5S RNA genes, was also defective in U6 snRNA transcription (Moenne et al. 1990). (6) Tagetitoxin, a specific inhibitor of transcription by pol III at low concentrations, inhibits the synthesis of U6 snRNA (Steinberg et al. 1990). All these data show that U6 snRNA genes are transcribed by pol III. Although the involvement of pol III in the synthesis of 7SK RNA is well documented (Zieve et al. 1977; Murph et al. 1986, 1987b; Kruger and Benecke 1987), the mepppG cap structure in 7SK RNA was only recently identified (Gupta et al. 1990b). It is significant to note that U3 snRNAs from tomato, pea and *Arabidopsis* do not contain the TMG cap structure and may contain the mepppA cap structure (Kiss and Solymosy 1990), and that U3 RNA in *Arabidopsis* is synthesized by pol III and not by pol II (Waibel et al. 1990). These observations are consistent with the notion that TMG-capped U snRNAs are pol II products and mepppG/A capped snRNAs are pol III products. All other known small nuclear RNAs, including RNaseP (Baer et al. 1990), add, MRP/7–2 (Hashimoto and Steitz 1983), and Alu-related B1, B2, and B3 RNAs (reviewed in Jelinek and Schmidt 1982), are synthesized by pol III.

4 Organization of U snRNA Genes

The organization and copy numbers of U snRNA genes vary from organism to organism. In general, the gene copy number correlates well with the abundance of each U snRNA; however, the copy number varies widely. In higher eukaryotes, most U snRNAs are represented by a multigene family. In lower eukaryotes, such as yeasts, most U snRNAs are coded for by single copy genes. The gene organization and copy number of U snRNA genes in different organisms is summarized in Table 3.

4.1 Human

Real genes have been characterized from the human genome for U1 (Manser and Gesteland 1981, 1982; Lund and Dahlberg 1984), U2 (VanArsdell and Weiner 1984; Westin et al. 1984), U3 (Suh et al. 1986; Yuan and Reddy 1988), U4 (Bark et al. 1986), and U6 snRNAs (Kunkel et al. 1986). Most, and perhaps all, of the human U1 snRNA genes are present as a tandem repeat on the short arm of the

Table 3. Copy number, organization and localization of U snRNA genes. (Slightly modified and updated from Dahlberg and Lund 1988)

Organism	RNA	Copy #[a]	Cloned genes	Organization	Location	Reference
Human	U1	~30	HSD1-7 HU1-1 cosD1, cosD21	Loosely clustered >44 kb apart, large tandem repeat unit	1p36	Manser and Gesteland (1981, 1982) Buckland et al. (1983) Lund and Dahlberg (1984) Bernstein et al. (1985)
	U2	10–20	U2.24A,B U2/6	Tightly clustered in one tandem array; 6.2 kb repeat unit	17q21 q22	Van Arsdell and Weiner (1984) Westin et al. (1984)
	U3	7–10	U3-1–4	If clustered, >10 kb apart	ND	Suh et al. (1986) Yuan and Reddy (1989)
	U4	100	U4C and U4B-like	Cloned genes tightly clustered	ND	Bark et al. (1986)
Rat	U6	ND	HU6	If clustered, >10 kb apart	ND	Kunkel et al. (1986)
	U1	~50	6-6A,B	Cloned genes 3.6 kb apart, in opposite orientation	ND	Watanabe-Nagasu et al. (1983)
	U2	40	RU2-3	If clustered, >10 kb apart	ND	Tani et al. (1983)
	U3	5–10	U3D, U3B.4,7 each type: 1-few	If clustered, >10 kb apart	ND	Stroke and Weiner (1985)
Mouse	U1	20–40 each type: 5–10	U1.1,2 (U1b2) U1a-236 (U1a1) U1b-136 (U1b2) U1b-453,550 (U1b6)	Inverted repeat, U1.1 and U1.2 5.0 kb apart; Genes for each type loosely clustered, more than 5–10 kb apart	3(U1b2,b3) 11 (U1a1) 12 (U1a2)	Marzluff et al. (1983) Blatt et al. (1988) Michael et al. (1986) Howard et al. (1986) Lund and Nesbitt (1988)
	U2	~10	U2 U2.47	One locus contains inverted repeat with 2 genes, 3.8 kb apart; Another locus one gene	ND	Nojima and Kornberg (1983) Moshier et al. (1987, 1988)

		Copy number	Clone	Organization	Location	References
	U3	6–7	U6–52	Two genes are 5kb apart	ND	Mazan and Bachelleri (1988)
	U6	2		If clustered, >9 kb apart	ND	Ohshima et al. (1981); Yuan and Reddy (1988)
Chicken	U1	6–10	U1.2.5	3 genes within 5kb; 1.8 kb apart	ND	Roop et al. (1981); Early et al. (1984)
	U2	35–40	U1–52a,b,c U2–6	Tightly clustered in tandem array, 5.35 kb apart	ND	Korf and Stumph (1988)
	U4	2	U4B,U4X	Cloned genes 465 bp apart	ND	Hoffman et al. (1986)
Xenopus	U1	Minor family (adult) ~50 Major family (embryonic) ~1000	X1U1.3 X1U1.8 X1U1b1, X1U1b2	Cloned locus has 3 genes within 5 kb Tightly clustered in large tandem array(s), 1.85 kb repeat unit with 1 copy each of the b1 and b2 genes	ND	Zeller et al. (1984); Mattaj and Zeller (1983); Lund et al. (1984); Krol et al. (1985); Ciliberto et al. (1985)
	U2	~500–1000	XLU2–5	Tightly clustered in large tandem array, 830 bp repeat unit	ND	Mattaj and Zeller (1983)
	U5	~100	XlU511H	Major family is a tandem repeat; 583 bp apart	ND	Kazmaier et al. (1987)
	U6	~600	XtU6–2	Tightly clustered in large tandem arrays, 1.6 and 1 kb repeat units		Krol et al. (1987)
Drosophila	U1	7	U14 DmU1.4 Dm6A	Dispersed	11B,21E 61A,82E 95C	Alonso et al. (1984b); Mount and Steitz (1981); Kejzlarova-Lepesant et al. (1984); Saluz et al. (1983, 1988)

Table 3. (*Continued*)

Organism	RNA	Copy #[a]	Cloned genes	Organization	Location	Reference
	U2	4–5	U2 131A, 131B U2 141A, 141B	Two unlinked clusters, each containing 2 genes within 3–5 kb	34BC 84C	Alonson et al. (1983, 1984a)
	U3	ND	DU3-1	Dispersed	ND	Akao et al. (1986)
	U4	3	U4-1, U4-2	Dispersed	39B, 40AB	Saba et al. (1986); Saluz et al. (1988)
	U5	7		Dispersed	14B,23D, 34A,35EF, 39B,63A	Saluz et al. (1988)
	U6	3	DU6–1, 6–2, 6–3	Closely linked, 500 bp apart, same orientation	96A	Das et al. (1987); Saluz et al. (1988)
Sea urchin	U1	20	LvU1.1 LvU1.2	Tightly clustered in large tandem arrays, 1.4 kb repeat unit; 2 types of repeat	ND	Brown et al. (1985) Nash and Marzluff (1988) Card et al. (1982) Yu et al. (1986)
	U2	20		1.1 kb repeat unit		Card et al. (1982)
	U7	5	U7	5 genes within 9.3 kb		Lorenzi et al. (1986)
Dictyostelium	U3	5	D2.1	Dispersed		Wise and Weiner (1980)
Trypanosome	U2	1	U2		ND	Tsuchidi et al. (1986); Mottram et al. (1989)
(T. brucei)	U4	1	U4		ND	Mottram et al. (1989)
	U6	1	U6		ND	Mottram et al. (1989)
C. elegans	U1	~11		Clusters of 2–3 genes or single copies dispersed in the genome	ND	Thomas et al. (1990)
	U2	~12			ND	Thomas et al. (1990)
	U4	~6			ND	Thomas et al. (1990)
	U5	~9			ND	Thomas et al. (1990)
	U6	~10			ND	Thomas et al. (1990)

Phaseolus	U1	1-few	U1	If more than one >14 kb apart	ND	van Santen and Spritz (1987)
Soybean	U1	2-4	U1a,b	Dispersed and tandem repeats		van Santen et al. (1988)
Tomato	U1	8 or more	U1.1 to U1.8	In 5 loci		Abel et al. (1989)
Arabidopsis	U2	10-15	U2.1 to U2.9	Do not appear to be clustered	ND	Vankan and Filipowicz (1988)
Maize	U2	25-40	U2-27			Brown and Waugh (1989)
Pea	U2	Many				Hanley and Schuler (1989)
Tomato	U3	1-few	U3	Genes and pseudogenes linked	ND	Kiss and Solymosy (1990)
Arabidopsis	U5	8-9	U5	Some gene linked	ND	Vankan et al. (1988)
Arabidopsis	U6		U6		ND	Waibel et al. (1990)
Tomato	U6		U6		ND	Szkukalek et al. (1990)
Yeast (*S. pombe*)	U2	1	pMa2		ND	Brennwald et al. (1988)
	U3	2		If linked, >8 kb apart	ND	Porter et al. (1988)
	U6	1			ND	Tani and Ohshima (1989)
Yeast (*S. cerevisiae*)	U1	1	SNR 19		ND	Siliciano et al. (1987a); Kretzner et al. (1987)
	U2	1	SNR20		ND	Ares (1986)
	U3	2	SNR17a,b	Genetically unlinked	ND	Hughes et al. (1987); Myslinski et al. (1990)
	U4	1	SNR3, 14		ND	Siliciano et al. (1987b)
	U5	1	SNR7		ND	Patterson and Guthrie (1987)
	U6	1	SNR6		ND	Brow and Guthrie (1988)
	U14	1	SNR128	67 bp apart from SNR190	ND	Zagorski et al. (1988)
	SNR3,4,5 8,9,10	1 each	SNR3,4,5,7, 8,9, 1sr1 (snR20)	Dispersed	ND	Tollervey et al. (1983); Wise et al. (1983) Parker et al. (1988)
	SNR 190	1	SNR190	67 bp apart from U14	ND	Zagorski et al. (1988)

[a] Copy # indicates the number of genes per haploid genome.

chromosome 1 (Lund et al. 1983). This result, obtained from human/rodent hybrid cells, was confirmed by in situ hybridization; these studies showed that U1 genes are clustered at 1p36 (Naylor et al. 1984; Lindgren et al. 1985b). The human U2 genes have been localized in 17q21–22 region (Lindgren et al. 1985a). Interestingly, the locations of U1 and U2 genes correspond to viral chromosome modification sites (Lindgren et al. 1985a). The flanking regions of up to 20 kb on either side of human U snRNA genes are highly conserved (e.g., Manser and Gesteland 1982; Bernstein et al. 1985). Since the pseudogenes for U snRNAs are abundant in the human genome (Denison et al. 1981), these conserved flanking regions were successfully used to estimate the true gene copy number in the midst of abundant pseudogenes (Lund and Dahlberg 1984).

4.2 Rodent

Rat U1 (Watanabe-Nagasu et al. 1983), U2 (Tani et al. 1983), U3 (Stroke and Weiner 1985), mouse U1 (Marzluff et al. 1983; Howard et al. 1986; Michael et al. 1986), U2 (Nojima and Kornberg 1983; Moshier et al. 1987), U3 (Mazan and Bachellerie 1988), and U6 (Ohshima et al. 1981) snRNA genes have been cloned and characterized. As in the case of human U snRNA genes, the flanking regions within each rodent U snRNA gene repeat are highly conserved; however, the 5′ flanking regions of different types of mouse U1 snRNA genes differ widely (Howard et al. 1986). Interestingly, the 5′ flanking sequences in the rat and mouse U1 gene repeats are the same (Moussa et al. 1987); similarly, the 5′ flanking sequences of the rat and mouse U3 snRNA genes are also conserved (Mazan and Bachellerie 1988). However, these flanking sequences differ from the 5′ flanking sequences in the corresponding human U1 or U3 snRNA genes. These data provide evidence for the conservation of snRNA gene repeats in closely related species.

4.3 Chicken

Genes for chicken U1 (Roop et al. 1981; Earley et al. 1984), U2 (Korf and Stumph 1986), and U4 (Hoffman et al. 1986) snRNAs have been isolated and characterized. The chicken U1 and U2 snRNA genes, are present as tandem repeats as in human and rodent genomes. The U1 and U2 snRNA genes are found in very different genomic environments but have similar promoter structures (Korf and Stumph 1986).

4.4 Xenopus

Frog U1 (Zeller et al. 1984; Lund et al. 1984; Krol et al. 1985; Ciliberto et al. 1985), U2 (Mattaj and Zeller 1983); U5 (Kazmaier et al. 1987), and U6 (Krol et al. 1987) snRNA genes have been characterized. The most unusual feature about the *Xenopus*

U snRNA genes is that the gene copy number is very high. More than 1000 copies each of U1 and U2 snRNA genes and about 600 copies of U6 snRNA genes are present in the *Xenopus* haploid genome. The abundance of these genes is similar to the observations made with 5S genes, of which over 20 000 copies are present in the *Xenopus* genome (reviewed in Long and Dawid 1980).

4.5 Drosophila

Drosophila U1 (Mount and Steitz 1981; Alonso et al. 1984b), U2 (Alonso et al. 1983; 1984a), U3 (Akao et al. 1986), U4 (Saba et al. 1986), and U6 (Das et al. 1987) snRNA genes have been characterized. Most of the U snRNA genes in the *Drosophila* genome have been mapped to particular chromosomal loci (Saluz et al. 1988). The genes for *Drosophila* U1, U4, and U5 snRNAs are dispersed, whereas genes for U2 snRNA are in two unlinked clusters. All the three genes for U6 snRNA are closely linked. While flanking sequences in the human U1 gene family are highly conserved, the flanking sequences of *Drosophila* U gene families are not well conserved (Alonso et al. 1984a; Das et al. 1987).

4.6 Sea Urchin

Several genes for U1 (Brown et al. 1985; Yu et al. 1986; Nash and Marzluff 1988), U2 (Card et al. 1982), and U7 (Lorenzi et al. 1986) have been isolated and characterized. There are multiple copies of genes for each U snRNA and these are tightly clustered. There are two types of U1 gene repeats and both types are transcribed in sea urchin embryos (Yu et al. 1986). Although U snRNA genes from human, rat and mouse are transcribed accurately in frog oocytes (see Sect. 6.2), sea urchin U7 snRNA genes are not expressed faithfully in frog oocytes (Strub and Birnstiel 1986)

4.7 Trypanosome

There has been interest in the structure of U snRNAs from *Trypanosomes* because of the *trans*-splicing and RNA editing that are common in these parasites (Simpson and Shaw 1989). Although several (U2, U4, and U6) capped snRNAs are found in *Trypanosomes* and are required for *trans*-splicing (Tsuchidi and Ullu 1990), they differ significantly from the metazoan U snRNAs (Tsuchidi et al. 1986; Mottram et al. 1989). For example, the Sm-binding site found in metazoan, yeast and plant U2 snRNAs is not present in *Trypanosomal* U2 snRNA. The analog for U1 snRNA has not yet been identified, and if U1 snRNA is even present in *Trypanosomes*, it appears to be a minor RNA or it lacks the TMG cap structure (Mottram et al. 1989). The spliced leader sequences contain TMG cap structure, associate with Sm-antigen, and serve as U1 snRNPs during the *trans*-splicing event (Bruznik et al. 1988; Thomas et al. 1988).

4.8 C. elegans

Nematodes are the only group of organisms in which both *cis*- and *trans*-splicing of nuclear mRNAs are known to occur. The genes for U1, U2, U4, U5, and U6 snRNAs from *C. elegans* have been isolated and characterized. The genes for each U snRNA is represented by a mutigene family and are dispersed randomly in the genome of *C. elegans* (Thomas et al. 1990).

4.9 Yeast

Many U snRNA genes, including U1 (Siliciano et al. 1987a; Kretzner et al. 1987), U2 (Ares 1986), U4 (Siliciano et al. 1987b), U5 (Patterson and Guthrie 1987), and U6 snRNA (Brow and Guthrie 1988), have been isolated from *S. cerevisiae*. With the exception of U3 snRNA genes (Hughes et al. 1987; Myslinski et al. 1990), all U snRNA genes that have been characterized from yeasts are single copy genes and are dispersed in the yeast genome. The U3 snRNA genes from some strains of *S. cerevisiae* contain introns (Myslinski et al. 1990). Genes for U2 (Brennwald et al. 1988). U3 (Porter et al. 1988), and U6 (Tani and Ohshima 1989) snRNAs have also been isolated from *S. pombe*. Interestingly, U6 snRNA genes in *S. pombe* (Tani and Ohshima 1989) and in several related fungi (Frendeway et al. 1990; Reich and Wise 1990) contain introns which resemble the introns found in yeast mRNAs. No pseudogenes have been reported for U snRNAs in the yeast genomes.

4.10 Plants

Genes coding for *Arabidopsis* U2 (Vankan and Filipowicz 1988), U5 (Vankan et al. 1988), and U6 (Waibel et al. 1990), bean U1 (van Santen and Spritz 1987; van Santen et al. 1988), tomato U1 (Abel et al. 1989), tomato U3 (Kiss and Solymosy 1990), U6 (Szkukalek et al. 1990), and maize U2 (Brown and Waugh 1989) snRNAs have been isolated and characterized. The plant U snRNA genes character-ized thus far are represented by multigene families and are not closely clustered. Most of the genes that have been isolated were shown to be real genes by the ex-pression of the cloned genes into electroporated plant protoplasts. Several pseudogenes for the plant snRNAs have been reported (e.g., bean U1 and tomato U3 snRNA pseudogenes); however, the pseudogenes do not appear to be as abun-dant as is the case in mammalian genomes.

4.11 Viral U RNAs

Recently, Lee et al. (1988) discovered that herpesvirus saimiri codes for at least five TMG-capped U snRNAs. These RNAs, in association with Sm antigen, are present as snRNP particles. All the herpesvirus U snRNA genes contain the consensus proximal sequence element, octamer, purine as an initiation nucleotide, and the 3'-end formation signals (Lee et al. 1988; Wasserman et al. 1989). These data suggest that herpesvirus U snRNAs and cellular TMG-capped U snRNAs are synthesized by the same cellular transcription machinery.

5 *Cis*-Acting Elements in U snRNA Genes

5.1 Initiation Nucleotide

Most U snRNAs thus far characterized initiate with a purine nucleotide. In this respect, the synthesis of U snRNAs is similar to that of mRNAs, tRNAs, and rRNAs which also initiate with a purine nucleotide. The TMG-capped U RNAs, in general, initiate with adenosine (Table 4); the only exceptions known are pea U5 and some yeast snRNAs, which initiate with guanosine (Krol et al. 1983; Parker et al. 1988). The U6 snRNA initiates with guanosine; interestingly, the initiation nucleotide of *Physarum* U6 snRNA is a pyrimidine (Skinner and Adams 1987). A consensus sequence 5' YYCAYYYY 3' was observed surrounding the cap site of mRNAs, wherein A is the initiation nucleotide (Corden et al. 1980). A pyrimidine-rich, conserved consensus sequence around the initiation site is also observed for the TMG-capped U snRNA genes (Table 4). Although fewer than ten genes have been characterized for U6 and 7SK snRNAs, a similar consensus (YYYGTNYT) is observed surrounding the initiation site. The sequences around the initiation nucleotide were found to be very important for both accuracy of initiation and the efficiency of transcription of *Xenopus* U6 snRNA gene (Mattaj et al. 1988).

5.2 Proximal Sequence Element (PSE)

The PSE of human U1 gene is absolutely required for transcription. This sequence lies approximately 45 to 70 bp upstream of the cap site, and is well conserved among U snRNA genes from the same species as well as between different species (reviewed in Dahlberg and Lund 1988). This was shown for the human U1 (Skuzeski et al. 1984; Murphy et al. 1987a), human U2 (Ares et al. 1985), chicken U4 (McNamara and Stumph 1990), frog U1 (Ciliberto et al. 1985), and frog U2 snRNA genes (Mattaj et al. 1985). In the PSE region, there is a 10–15 bp-long sequence conserved among vertebrate U snRNA genes (Table 4). This conserved region is able to bind a factor found in nuclear extracts (Gunderson et al. 1988; Tebb et al. 1987). A protein heterodimer (63/80 kDa) has been purified to homogeneity by PSE DNA affinity chromatography (Roberts et al. 1989; Knuth et al. 1990). This factor was required for transcription in vitro of human transferrin receptor gene promoter, but not for transcription of adenovirus 2 major late promoter; its role in transcription of snRNA promoters remains to be established (Knuth et al. 1990). This protein heterodimer is similar, perhaps identical, to the Ku protein; interestingly, many patients with autoimmune diseases produce antibodies against Ku protein (re-

Table 4. *Cis*-acting elements in U snRNA genes

A. Consensus sequences surrounding the initiation site for TMG-capped snRNAs

Nucleotide no.		-3	-2	-1	1	$+1$	$+2$	$+3$	$+4$
Nucleotide	G	17	23	1	3	6	9	11	0
	A	24	3	1	59	9	27	11	5
	T	11	28	19	0	45	1	8	35
	C	10	8	41	0	2	25	32	22
Consensus		N	G/T	Y	A	T	A/C	C	Y

B. Consensus PSE sequences

	G G T C
Human	CTCACCGTGAGT(GT)RAAR$_{0-3}$TG
(Dalhberg and Lund 1988)	
	T
X. laevis (Parry et al. 1989)	CTCTCCNYRAG
Plant (Vankan et al. 1988)	RTCCCACATCG

C. Consensus 3′-end formation signal

	A
Animal snRNA genes (Hernandez 1985)	GTTN$_{0-3}$AAAGNNAGA
(Yuo et al. 1985)	
	T
Plant snRNA genes (Vankan et al. 1988)	AGTNAAA

A. The consensus sequence is derived from 62 snRNA genes coding for TMG-capped U snRNAs. These genes are listed in Table 3.
B. The sequence in brackets is not present in all genes, Y, pyrimidine, R, purine, N, any of the four nucleotides.

viewed in Tan 1989). In addition to its involvement in transcription initiation, the PSE also plays a role in 3′ end formation (see Sect. 5.5). Since the PSE sequence fixes the transcription initiation site, the PSE is also referred to as the snRNA TATA box (reviewed in Dahlberg and Lund 1988). However, the PSE differs from the conventional TATA boxes in both location and sequence. Vertebrate U6 snRNA genes and plant U snRNA genes contain a functional PSE around -50, and in addition, a conventional TATA box at -30 position.

5.3 TATA Box

A TATA box was first observed at -30 position in a mouse U6 snRNA gene (Ohshima et al. 1981) and a TATA box in the same position is found in all plant U snRNA genes (e.g., Vankan and Filipowicz 1989). Instead of at the usual -30 position, several yeast U snRNA genes contain a TATA box at -92 position (Parker et al. 1988); the functional significance of this unusual positioning of these TATA boxes is not known. The TATA box has been shown to play important roles in transcription of U6 snRNA genes (Carbon et al. 1987; Kunkel and Pederson 1989) and is also required for transcription of plant U snRNA genes (Vankan and Filipowicz

1989). It is not clear whether the TATA box-binding factor TFIID, characterized for the mRNA genes (Horikoshi et al. 1989), is shared by the U snRNA genes. In the case of *S. cerevisiae*, it is suggested that TFIIIB may be the factor that interacts with the U6 gene TATA box (Brow and Guthrie 1990).

5.4 Distal Sequence Element (DSE)

Although U snRNA genes containing only 80 bp of upstream sequences support transcription, sequences further upstream enhance the transcription up to 50-fold and are also required for the formation of a stable transcription complex (Skuzeski et al. 1984; Ciliberto et al. 1985; Ares et al. 1985; Mangin et al. 1986; Mattaj et al. 1985; Krol et al. 1985; Murphy et al. 1987a). These distal sequences (DSE), consisting of discrete conserved sequence motifs, normally located around 250 bp upstream of the transcription start site, are collectively termed the DSE. In all DSEs studied, there is at least one copy of the octamer motif, ATGCAAAT. This octamer motif is found in combination with a variety of other transcription factor-binding sites which differ in different DSEs. These sequence motifs include SP1, AP-2, SPH, cAMP response element binding protein (CREB) and are similar to the sequence motifs present in the mRNA promoters (Table 5). In some cases, but not all, the DSEs can be productively exchanged between different U RNA genes and mRNA genes. For example, the DSE of the human U2 gene can be functionally replaced by SV40 enhancer (Mangin et al. 1986) and DSEs between U2 and U6 snRNAs are interchangeable (Bark et al. 1987). In some cases, the DSEs were not interchangeable between mRNA and snRNA genes (Dahlberg and Schenborn 1988; Tanaka et al. 1988). *Xenopus* U2 DSE can functionally replace the SV40 enhancer to allow

Table 5. Sequence motifs and *trans*-acting factors involved in the transcription of U snRNA genes

Cis-element	*Trans*-acting factor	U gene	Reference
ATGCAAAT	Oct-1	Human U1	Mangin et al. (1986)
GGGCGGGGC	SP1	Human U2	Manging et al. (1986)
TCCCCAGCGTCCCAAG	AP-2	Human U4	Weller et al. (1988)
TGACGTCA	CRE	Human U4	Weller et al. (1988)
TATAA	TFIIIB	Mouse U6	Brow and Guthrie (1990)
GGGTCCGGG	MSE	*Xenopus* U2	Tebb et al. (1987)
GTGGCAGTC	D2	*Xenopus* U2	Tebb and Mattaj (1989)
GGTGCGCGCCGG	D5	*Xenopus* U5	Kazmaier et al. (1987)
CGCGCGCTGCA-TGCCGGGAG	SPH	Chicken U1	Roebuck et al. (1987)
GTGACCGTGTG-TGTAAAGAGTG	Ku protein	Human U1	Skuzeski et al. (1984); Knuth et al. (1990)

Only one representative gene and reference is given, although many references are available in the literature (see Dahlberg and Lund 1988; Parry et al. 1989; Kleinert et al. 1990).

replication and T antigen expression in SV40 mutants lacking the SV40 enhancer (Mattaj et al. 1985).

The action of DSE is independent of orientation and, to a limited extent, independent of its position upstream of the cap site. Some DSEs also function when placed downstream of the U snRNA genes (Mangin et al. 1986; Ares et al. 1987). Interactions between the factors binding to the core promoter and the DSE have been studied by several groups and various models have been proposed (Tanaka et al. 1988; Roebuck et al. 1990; Parry and Mattaj 1990). Purified octamer binding factors (e.g., OTF-1 and OTF-2), involved in stimulating the transcription by pol II, stimulate pol III-mediated transcription of a human 7SK gene (Murphy et al. 1989); however, octamer-motif in 7SK RNA gene can be deleted without affecting transcription in vivo (Kleinert et al. 1990). Cooperative interactions between transcription factors Sp1 and octamer-binding factor (OTF-1) were shown to be required for optimal transcription of human U2 snRNA gene in vivo (Janson and Pettersson 1990). These data indicate that octamer-binding factors are involved in the transcription of mRNA, TMG-capped and mepppG-capped snRNAs.

While many factors are known to be involved in the transcription of U snRNA genes, it is not clear which one(s) serve as the transcription factor and which serve as assembly or auxiliary factors. In the case of 5S RNA gene transcription. TFIIIB appears to be the transcription factor while TFIIIA and TFIIIC are assembly factors. Once the TFIIIB is properly assembled onto 5S DNA, TFIIIA and TFIIIC are no longer required (Kassavetis et al. 1990). A similar mechanism appears to be operative in the transcription of rDNA genes by pol I (Paule 1990). Although the PSE is an important and required element in the transcription of U snRNA genes, some transcription of U6 snRNA gene is observed in DNA constructs lacking the PSE but containing the B box which binds TFIIIC and hence perhaps can position TFIIIB (Parry and Mattaj 1990). Therefore, it is possible that TFIIIB, or a factor related to TFIIIB, is the transcription factor for U6 snRNA genes.

5.5 3′-End Formation

The 3′ ends of most mRNAs are formed by endonucleolytic cleavage of a long precursor RNA followed by polyadenylation. The 3′-end formation of TMG-capped U RNAs is different and rather complex. There is a required cis-element located 10–20 bp downstream of the 3′ end of mature U snRNAs (Yuo et al. 1985; Hernandez 1985; Neuman de Vegvar et al. 1986). A consensus sequence is present in this region (Hernandez 1985; Yuo et al. 1985; Vankan et al. 1989; reviewed in Dahlberg and Lund 1988; Table 4); however, the sequences in this highly conserved and required 3′-end formation signal are quite tolerant to mutation (Ach and Weiner 1987). Another unusual feature is that the 3′-end formation is coupled to transcription from snRNA promoters (Hernandez and Weiner 1986; Neuman deVegvar et al. 1986). Specifically, the PSE in an snRNA promoter appears to confer this specificity to the snRNA gene transcription complexes, enabling them to recognize the 3′-end formation signal (Hernandez and Lucito 1988; Neuman de Vegvar and

Dahlberg 1989; Parry et al.1989). However *S. cerevisiae* U5 snRNA, with the correct 3' ends, is expressed from yeast GAL1 promoter suggesting that, unlike the situation in higher eukaryotes, proper 3'-end generation of yeast U5 RNA does not require specialized initiation events (Patterson and Guthrie 1987).

Like mRNAs, the primary transcripts of the TMG-capped snRNAs extend beyond the 3'-end of the mature RNAs (Eliceiri and Sayavedra 1976). These short-lived precursor snRNAs are up to 12 nucleotides longer than the mature snRNAs, and are processed in the cytoplasm (Madore et al. 1984a,b; Zieve et al. 1988; Neuman de Vegvar 1990; reviewed in Dahlberg and Lund 1988; Zieve and Sauterer 1990). Recently, an accurate and efficient in vitro processing system for 3'-end maturation of U2 snRNA has been developed (Kleinschmidt and Pederson 1987).

The 3'-end formation of mepppG-capped U6 and 7SK RNAs appears to be similar, if not identical to the 3'-end formation of eukaryotic 5S RNA. A cluster of four or more T residues is found in both U6 and 7SK genes corresponding to their 3' ends. Removal of this T-cluster in the mouse U6 snRNA gene results in read-through transcription (Das et al. 1988). The plant U3 genes transcribed by pol III also contain a cluster of T residues on their 3' ends (Kiss and Solymosy 1990; Waibel et al. 1990). It seems likely that the DNA sequences and *trans*-acting factors governing the 3'-end formation of mepppG/A-capped RNAs are the same as those for other previously characterized pol III genes (Bogenhagen and Brown 1981; Geiduschek and Tocchini-Valentini 1988).

6 Interconversion of U snRNA Promoters

The 5' flanking sequences of vertebrate U1, U2 or U6 snRNA genes are capable of directing transcription, indicating that the promoters for these U snRNA genes may be exclusively in the 5' flanking region. When one compares the promoters of vertebrate U1 and U2 genes with those of the U6 genes, the main difference is the presence of a TATA box at position −30 in the U6 genes and its absence in U1 and U2 snRNA genes (Fig. 2). Therefore, it should be possible to convert the U1, U2 (pol II) promoters into U6 (pol III) promoters and vice versa. This has been successfully accomplished in the case of *Xenopus* and human U snRNA genes (Mattaj et al. 1988; Lobo and Hernandez 1989; Lobo et al. 1990). Alteration in the −30 TATA box in the human U6 gene results in transcription by pol II and introduction of a TATA box into the −30 region in the human U2 gene results in transcription by pol III (Lobo and Hernandez 1989). In the case of plants, genes for TMG-capped U2, U4, and U5 snRNAs are transcribed by pol II and genes for mepppG-capped U6 snRNA are transcribed by pol III. Both classes of genes require only two *cis*-acting elements which are interchangeable. The spacing between the PSE and the TATA box appears to be a major factor in determining which polymerase is employed. Introduction of a 10 bp DNA between the PSE and the TATA box of *Arabidopsis* U6 gene results in transcription by pol II. Similarly, deletion of 10 bp between the TATA box and the PSE of *Arabidopsis* U2 gene results in transcription by pol III

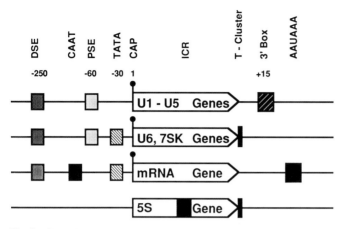

Fig. 2. *Cis*-acting elements in pol II and pol III genes. Diagrammatic representation of some functional cis-acting elements found in TMG-capped snRNA genes (e.g., U1–U5), mepppG-capped snRNA genes (e.g., U6, 7SK), eukaryotic mRNA gene (e.g., β-globin) and mammalian 5S RNA gene. ICR, internal control region. The position of DSE in mRNA genes is much more variable than in U snRNA genes. Some 5S RNA genes contain and require upstream TATA-like sequences. (Tyler 1985)

(Waibel et al. 1990). These data provide strong support for a common evolutionary origin of RNA polymerases. In fact, it is possible that some U snRNA promoters, such as the *Xenopus* U6 snRNA gene promoter, may have dual specificity. Microinjected *Xenopus* U6 snRNA genes apparently are transcribed in frog oocytes by both pol II and pol III (Mattaj et al. 1988).

6.1 SnRNA and mRNA Gene Promoters Are Distinct

Although snRNA gene promoters resemble the promoters of mRNA genes, there are many differences. None of the snRNAs characterized thus far is polyadenylated. Transcripts initiated at an eukaryotic U1 snRNA promoter are not polyadenylated, even when the snRNA coding region is deleted and normal pre-mRNA cleavage and polyadenylation signals are included (Neuman deVegvar et al. 1986). The U1 snRNA promoter and enhancer do not direct synthesis of messenger RNA. When fused to a chloramphenicol acetyl transferase (CAT) coding region, the upstream sequences of the human U1 gene were able to stimulate the synthesis of functional CAT mRNA. However, the polyadenylated CAT mRNA is initiated at cryptic sites upstream of the U1 sequences. The RNA initiated at the normal +1 position (in relation to U1 gene promoter) was not polyadenylated (Dahlberg and Schenborn 1988). These results demonstrate that the U1 snRNA gene promoter, although very efficient in the transcription of snRNAs, is unable to transcribe polyadenylated mRNAs. When HeLa cells are exposed to ultraviolet light, synthesis of U1 snRNA is inhibited more than the synthesis of 5S RNA or mRNAs (Eliceiri and Smith

1983; Morra et al. 1986). Since the U snRNA genes are small in size, the unusual sensitivity of snRNA transcription to UV light is surprising. All these data show that mRNA gene transcription complexes are distinct from human U1 gene transcription complexes.

6.2 Transcription of snRNA Genes in Vitro and in Heterologous Systems

TMG-capped U snRNA genes can be transcribed, to a limited extent, across species. Human, rat, mouse, and chicken snRNA genes are transcribed in frog oocytes (Murphy et al. 1982; Skuzeski et al. 1984; Nojima and Kornberg 1983; Howard et al. 1986; Hoffman et al. 1986; Reddy et al. 1987); however, the efficiency of this heterologous transcription is less than the efficiency of homologous transcription of *Xenopus* U snRNA genes in frog oocytes. The *Drosophila* U4 snRNA genes or the sea urchin U7 snRNA genes are not transcribed in frog oocytes (Saba et al. 1986; Birnstiel and Schaufele 1988). In contrast, the U6 snRNA genes are transcribed in evolutionarily distant species. U6 snRNA gene from *S. pombe*, *Drosophila*, *Xenopus*, and mouse are transcribed in HeLa cell extracts in vitro (Kleinschmidt et al. 1990; Singh et al. 1990).

7 Regulation of snRNA Synthesis

In many organisms, there are several structural variants for U snRNAs; these structural variants differ from each other from <1% up to 15% in primary sequence. Structural variants were first reported for mouse U1 snRNA (Lerner and Steitz 1979) and rat U3 snRNA (Reddy et al. 1979). The structural variants are usually designated U RNA followed by A,B,C, etc. For example, the two major classes of U1 variants are termed U1A and U1B (Lund et al. 1985; Kato and Harada 1985; Lund 1988; reviewed in Dahlberg and Lund 1988); the rat U3 variants are termed U3A, U3B, and U3C. All of these U snRNA variants differ from each other slightly in their primary sequences, but these minor differences do not alter their ability to assume similar secondary structures. These variant snRNAs may be responsible for tissue-specific and/or stage-specific processing of pre-mRNAs. Alternations in the sequence of U1 snRNA have been shown to induce alternative splicing of precursor mRNAs (Yuo and Weiner 1989). Differential expression of U snRNAs between embryonic and adult cells has been observed in mouse (Lund et al. 1985), *Xenopus* (Lund and Dahlberg 1987), Chicken (Korf and Stumph 1989), and sea urchin (Nash et al. 1989) cells. *Xenopus* is unusual in that it contains large number of genes for U snRNAs (Table 3). Among these U snRNA genes, embryonic and adult genes have both been distinguished in the case of U1, U2, and U4 genes. In general, the U1a (adult) snRNAs are synthesized in all transcriptionally active cells and U1b (embryonic) snRNA accumulates only in cells that are capable of further differentiation (summarized in Dahlberg and Lund 1988). Similar developmental regulation of two

U4 snRNA genes have been demonstrated in different adult and embryonic chicken tissues (Korf et al. 1988).

There is another type of developmental regulation that is observed for U snRNA genes in *Xenopus*. During early embryonic development there is little synthesis of any type of RNA. At the mid-blastula transition, the endogenous or injected *Xenopus* U1 genes are strongly activated but the injected human U1 genes are not transcribed, providing further evidence for species-specific and/or gene-specific transcription factors (Dahlberg and Lund 1988). SnRNA transcription is also regulated during oogenesis and egg maturation (Forbes et al. 1983, 1984; Fritz et al. 1984; Lund and Dahlberg 1987; Lund et al. 1987; Dahlberg and Lund 1988).

8 Formation of Cap Structure in U snRNAs

There are two types of cap structures found in the U series of snRNAs (Fig. 1). The TMG cap structure is found in U1-U5 and U7-U14, and several other snRNAs (Reddy and Busch 1988). The TMG cap structure has been demonstrated in U snRNAs from diverse organisms including human, yeast, *Trypanosome*, and plant cells. The anti-TMG antibodies immunoprecipitate snRNAs and snRNPs from diverse eukaryotic cells (Bringmann and Luhrmann 1986; Krol et al. 1983; Montzka and Steitz 1988; Riedel et al. 1986). In the case of TMG-capped U snRNAs the m^7G cap is added co-transcriptionally in the nucleus (Eliceiri 1980; Skuzeski et al. 1984); this m^7G cap is converted to a TMG cap in the cytoplasm, and this trimethylation is dependent on an Sm-binding site in some snRNAs (Fig. 3). The position of the Sm-binding site in relation to the cap site is not critical (Mattaj 1986).

The mepppG cap structure has been shown, only in human U6 and 7SK snRNAs (Singh and Reddy 1989; Gupta et al. 1990b). However, the 5' end of U6 snRNAs from rat, mouse, *Trypanosomes, Physarum*, plant, and dinoflagellate cells were earlier shown to contain a cap structure different from TMG cap (Singh and Reddy 1989; Mottram et al. 1989). Antibodies specific to mepppG are able to immunoprecipitate U6 snRNA from human, rat, and plant RNAs (Gupta et al. 1990b) and *Xenopus* RNAs (E. Lund pers. commun.). In addition, U6 RNA is the most conserved of all the U snRNAs (Das et al. 1987; Brow and Guthrie 1988). Therefore, it appears that the U6 snRNAs from other species also have the mepppG cap structure as shown in Fig. 1.

8.1 Signal for U6 snRNA Capping

While the TMG-capped RNAs are transcribed by RNA polymerase II (Dahlberg and Lund 1988), the mepppG cap-containing U6 and 7SK RNAs are transcribed by RNA polymerase III (Table 2). The capping of the RNAs transcribed by pol II is coupled to transcription, and the capping of these RNAs does not require any specific RNA sequence (reviewed in Banerjee 1980). Experiments conducted to delin-

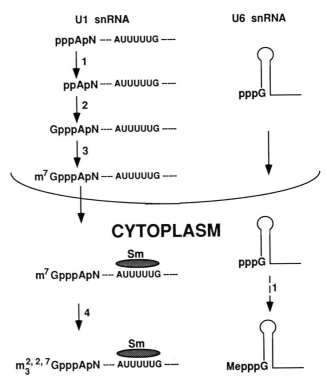

Fig. 3. Mechanism of snRNA cap formation. Schematic representation of the cellular events leading to the formation of snRNA cap structures. TMG-capped snRNAs (e.g., U1 snRNA): *1* triphosphatase; *2* guanylyltransferase requiring GTP; *3* methyltransferase; *4* Sm-dependent methyltransferase. MepppG-capped snRNAs (e.g., U6 snRNA): *1* RNA sequence-dependent methyltransferase. The methyltransferases utilize S-adenosyl methionine as the methyl donor

Table 6. Similarities and differences in the capping of TMGpppA- and mepppG/A-capped RNAs

Property	TMG-capped RNAs	mepppG-capped
Cap structure	TMGpppA/G	mepppG/A
Examples	U1–U5, U7–U14	U6, 7SK
RNA polymerase	Pol II (B)	Pol III (C)
Sequence-dependent	No	Yes
Coupled to transcription	Yes	No
Methyl group donor	S-adenosyl methionine	S-adenosyl methionine

eate the signal for the capping of U6 snRNA showed that the transcripts containing the entire U6 snRNA or only the sequences corresponding to nucleotides 1–25 of U6 snRNA were efficiently capped. The transcripts lacking nucleotides 1–25 of U6 snRNA were not capped, showing that nucleotides 1–25 of U6 snRNA are necessary and sufficient for optimal capping in vitro, and the information necessary for capping lies within the U6 snRNA, rather than in the transcription complex of the U6 snRNA gene (Singh et al. 1990).

The 5′ ends of known U6 RNAs from plants, yeast, or vertebrates can form a stem-loop structure, even though the length and nucleotide composition of the stem-loop differs among organisms (Epstein et al. 1980; Mottram et al. 1989; Roiha et al. 1989; Tani and Ohshima 1989). In mammalian U6 snRNA, the stem-loop is followed by an AUAUAC sequence in the single-stranded region. Mutagenesis of either the AUAUAC or the conserved stem-loop structure decreased capping efficiency. Therefore, the phylogenetically conserved stem-loop structure and the AUAUAC sequence are necessary for wild-type capping efficiency (Singh et al. 1990). Since the capping determinant appears to be rather simple, it should be possible to convert a noncapped RNA into an RNA suitable for U6 snRNA-specific capping. Transcripts corresponding to the vector DNA which do not get capped were capped when an AUAUAC sequence was introduced following a synthetic stem-loop structure; the efficiency of capping, however, was less than that of wild-type U6 snRNA. These data indicate that an unrelated stem-loop followed by an AUAUAC sequence is sufficient to direct capping in vitro. However, specific sequences within the stem-loop also play a role in the capping of U6 snRNA since not all stem-loops are substrates for U6 snRNA-specific capping (our unpublished results). U6 snRNAs synthesized from *Xenopus* or *Drosophila* U6 genes were efficiently and accurately capped in HeLa cell extracts. In addition, transcripts containing the 5′ portion of *S. cerevisiae* U6 snRNA were also accurately capped in HeLa cell extracts (Singh et al. 1990), showing evolutionary conservation of the capping machinery for mepppG-capped snRNAs.

8.2 Post-Transcriptional Capping of U6 snRNA

Several lines of evidence suggest that transcription and capping of mRNAs and U1-U5 snRNAs are concurrent. Studies with viral and cellular mRNAs (Furuichi 1978; reviewed in Banerjee 1980), as well as with U1-U5 snRNAs (Skuzeski et al. 1984; Mattaj 1986), showed that the capping and transcription by pol II are coupled. Furthermore, over 95% of the transcripts synthesized by heterologous T7 RNA polymerase in mammalian cells remain uncapped, presumably because of the inability of the capping machinery to interact with heterologous T7 RNA polymerase (Fuerst and Moss 1989). These data confirm that the capping of RNAs transcribed by pol II is tightly linked to transcription.

However, noncapped U6 snRNA synthesized in vitro with T7 polymerase was accurately capped to mepppG when incubated with the HeLa cell extracts, indicating that the transcription and capping of the U6 snRNA can be uncoupled. Also, in frog oocytes injected with mouse or *Xenopus* U6 snRNA genes, a large proportion of the transcripts are not capped (E. Lund pers. commun.). When T7-U6 snRNA was injected into enucleated frog oocytes, accurate capping of U6 snRNA was observed (our unpublished results), suggesting that the capping of U6 snRNA can occur post-transcriptionally in the cytoplasm (Fig. 3). These data, however, do not rule out the possibility of capping occurring in the nuclear compartment. During the capping of mRNAs or TMG capped U snRNAs, the γ-phosphate is removed (Fig. 3); however, during the capping of U6 snRNA, the γ-phosphate is retained in the U6 snRNA cap. In addition, the U6 snRNAs containing pG or ppG on their 5′ ends are not recognized as substrates for methylation, indicating that the capping machinery of U6 snRNA is highly specific for the γ-phosphate of the initiation nucleotide (Gupta et al. 1990a).

In studies with three different T7 RNA polymerase-transcribed U6 snRNAs, in which the pppG was shifted 32, 51, or 361 nucleotides away with reference to the capping determinant, there was no readily detectable cap structure formed. These data show that in the case of U6 snRNA the γ-phosphate must reside in close proximity to the capping determinant for wild-type capping efficiency (Singh et al. 1990; Gupta et al. 1990a). The 5′ 29 nts of human 7SK RNA were sufficient to direct cap formation in vitro (our unpublished results). These results suggest a common underlying mechanism in the capping of mepppG-capped snRNAs. The capping of U6 snRNA requires at least one heat-labile factor, presumably a protein, and S-adenosylmethionine as a methyl group donor. The capping is inhibited by the addition of S-adenosylhomocysteine, a known inhibitor of methylation reactions mediated by S-adenosylmethionine (Gupta et al. 1990a). The similarities and differences in the capping of TMG- and meppp-capped snRNAs are summarized in Table 6.

8.3 Other snRNAs with mepppG Cap Structure

Although there are only two snRNAs known to be capped with the mepppG cap structure, there are data to suggest that many other cellular small RNAs are capped with mepppG. When labeled HeLa cell RNAs were immunoprecipitated with anti-mepppG antibodies, many small RNAs ranging in length from approximately 50 to 330 nucleotides were found in the immunoprecipitates (Gupta et al. 1990b), suggesting that a wide spectrum of HeLa cell small RNAs contain the mepppG cap structure. The concentrations of these small RNAs are very low (<10 000 copies/cell).

Characterization and studies on these minor RNAs and their genes will provide more insight into the cellular mechanisms involved in the synthesis of this new class of U snRNAs.

8.4 Functions of Cap Structures

The cap structure of mRNAs has been shown to enhance the stability of mRNAs by protecting them against 5' exonucleolytic degradation and to increase the translational efficiency by facilitating the formation of an initiation complex (reviewed in Banerjee 1980; Shatkin 1985). In addition, recent studies indicate that the cap structure plays important roles in mRNA biogenesis. These roles include transcription initiation (Shatkin 1976; Furuichi 1978), generation of capped primers necessary for viral mRNA synthesis (Ulmanen et al. 1983), pre-mRNA splicing (Konarska et al. 1984; Edery and Sonenberg 1985), and 3' processing of mRNAs (Georgiev et al. 1984; Hart et al. 1985).

RNAs of cowpea mosaic virus and three picornaviruses have been shown to bear a genome-associated protein, which is covalently linked to the 5' end of the RNA (reviewed in Banerjee 1980). In the case of poliovirus RNA, a specific protein, VPg, with U residues attached to it, may serve as a primer during RNA replication (Flanegan et al. 1977). The function(s) of TMG cap structure in other U snRNAs and of meppppG cap structure in U6 and 7SK snRNAs are not known. In the case of U2 snRNA, the TMG cap does not appear to be absolutely required for 3'-end processing (Hernandez and Weiner 1986), small nuclear ribonucleoprotein assembly, and/or the targeting of the U2 snRNA to nucleus (Kleinschmidt and Pederson 1990).

Using the frog oocyte system, two groups showed that the TMG cap in the case of U1 and U2 snRNAs is required for transport of these RNPs from cytoplasm to the nucleus (Fisher and Luhrmann 1990; Hamm et al. 1990). SnRNAs with m^7G or $m^{2,7}G$ or m^7A caps were not transported. The nuclear targeting of U6 snRNA requires the AUAUAC sequence (nucleotides 20–25) (Hamm and Mattaj 1989) and probably also the U6 snRNA cap structure, γ-monomethyl phosphate (Hamm et al. 1990); this AUAUAC sequence is essential for the capping of U6 snRNA (Singh et al. 1990). The nuclear targeting of U1 snRNA requires both the TMG cap and binding of common U snRNP protein (Sm-protein). However, the interaction between TMG cap structure and Sm-protein is not essential, since the TMG cap can functionally substitute the methyl cap of U6 snRNA for the nuclear targeting of U6 RNA as long as the AUAUAC sequence is present (Hamm et al. 1990). These data suggest that the two cap structures may be functionally equivalent, as are the Sm-binding site in U1 snRNA and the AUAUAC sequence in U6 snRNA. However, the AUAUAC sequence alone may not be sufficient for nuclear targeting; it is likely that the whole capping determinant (stem-loop and AUAUAC sequence) plays a role in nuclear targeting.

The capped U1 and U2 snRNAs were more stable than the noncapped RNA, indicating that the TMG cap increases the stability of U2 snRNA (Kleinschmidt and Pederson 1989; Hamm et al. 1990) and this may hold true for other TMG cap-containing U snRNAs as well. The methylation of the γ-phosphate may protect U6 RNA from exonucleolytic degradation. In fact, while methylguanosine cap structures can be cleaved from capped RNAs by snake venom phosphodiesterase, the U6 cap is resistant (Epstein et al. 1980). In U6 snRNA, deletion of the nucleotides 1–52 did not inhibit the assembly of U6 RNA into the U4/U6 snRNP and deletion of nucleotides 1–37 did not affect spliceosome assembly (Bindereif et al. 1990). Since these deletions encompass the capping determinant (nucleotides 1–25) of U6 snRNA, the U6 snRNA cap structure may not play a direct role in the assembly of U4/U6 snRNP or of the spliceosome. The U3 snRNA from the animal cells contains a TMG cap and apparently an mepppA cap in higher plant cells (Kiss and Solymosy 1990). Therefore, it is likely that the TMG cap and the mepppG cap, at least in the case of U3 snRNA, are functionally interchangeable.

9 Summary

The organization of U snRNA genes in higher eukaryotes is very similar to that of genes for ribosomal RNAs; snRNA genes are present in multiple copies arranged in tandem repeats. Introns are usually not present in the U snRNA genes; however, in two instances where introns are found, they resemble the mRNA introns in pre-mRNAs. Therefore, the U snRNAs required for the splicing of pre-mRNAs apparently also participate in the splicing of their own introns. The U snRNA genes can be classified into two classes depending on the cap structure on their 5′ ends. The TMG-capped snRNA genes are transcribed by pol II, whereas mepppG-capped snRNAs are transcribed by pol III. Promoters for U snRNA genes are unique and distinct from promoters for other genes. Both pol III and pol II apparently use many common transcription factors in transcribing U snRNA genes. All the U snRNAs are rapidly transported into the cytoplasm where they are modified and assembled into ribonucleoprotein particles. One of the modifications that occurs in the cytoplasm is the capping of mepppG-capped snRNAs. The snRNPs are then transported into the nucleus where they are used in the processing of precursor RNAs. The TMG and mepppG cap structures appear to be required for the transport of these RNAs from the cytoplasm to the nucleus.

Abbreviations
DSE, distal sequence element
mepppG, γ-monomethyl guanosine triphosphate
pol II, RNA polymerase II
pol III, RNA polymerase III
PSE, proximal sequence element
snRNA, small nuclear RNA
TMG, trimethylguanosine
U snRNA, uridylic acid-rich small nuclear RNA

Acknowledgements. The studies from our lab are supported by grants from the U.S. Dept. of Health and Human Services (GM13120 and CA10893). We thank Harris Busch for unflagging optimism and constant support, James Dahlberg, Elsebet Lund, and Garry Zieve for valuable suggestions, and Shashi Gupta, Dale Henning, Laura Leo, and Ravindra Reddy for careful reading of the manuscript.

References

Abel S, Kiss T, Solymosy F (1989) Molecular analysis of eight U1 gene candidates from tomato that could potentially be transcribed into U1 RNA sequence variants differing from each other in similar regions of secondary structure. Nucl Acids Res 17:6319–6337

Ach RA, Weiner AM (1987) The highly conserved U small nuclear RNA 3'-end formation signal is quite tolerant to mutation. Mol Cell Biol 7:2070–2079

Akao M, Reddy R, Busch H (1986) Multiple sequences in the *Drosophila melanogaster* U3 RNA gene are homologous to vertebrate U3 RNA. Biochem Biophys Res Commun 138:512–518

Alonso A, Jorcano JL, Beck E, Spiess E (1983) Isolation and characterization of *Drosophila melanogaster* U2 small nuclear RNA genes. J Mol Biol 169:691–705

Alonso A, Beck E, Jorcano JL, Hovemann B (1984a) Divergence of U2 snRNA sequences in the gemone of *D. melanogaster*. Nucl Acids Res 12:9543–9550

Alonso A, Jorcano JL, Beck E, Hovemann B, Schmidt T (1984b) *Drosophila melanogaster* U1 snRNA genes. J Mol Biol 180:825–836

Ares M Jr (1986) U2 RNA from yeast is unexpectedly large and contains homology to vertebrate U4, U5, and U6 small nuclear RNAs. Cell 47:49–59

Ares M Jr, Mangin M, Weiner AM (1985) Orientation-dependent transcriptional activator upstream of a human U2 snRNA gene. Mol Cell Biol 5:1560–1570

Ares M Jr, Chung JS, Giglio L, Weiner AM (1987) Distinct factors with Sp1 and NF-A specificities bind to adjacent functional elements of the human U2 snRNA gene enhancer. Genes Develop 1:808–817

Baer M, Nilse TW, Costigan C, Altman S (1990) Structure and transcription of a human gene for H1 RNA, the RNA component of human RNaseP. Nucl Acids Res 18:97–103

Banerjee A (1980) 5'-terminal cap structure in eukaryotic messenger ribonucleic acids. Microbiol Rev 44:175–205

Bark C, Weller P, Zabielski J, Pettersson U (1986) Genes for human U4 small nuclear RNA. Gene 50:333–344

Bark C, Weller P, Zabielski J, Janson L, Pettersson U (1987) A distant enhancer element is required for polymerase III transcription of a U6 RNA gene. Nature (Lond) 328:356–359

Berget SM, Robberson BL (1986) U1, U2, and U4/U6 small nuclear ribonucleoproteins are required for in vitro splicing but not polyadenylation. Cell 46:691–696

Bernstein LB, Manser T, Weiner AM (1985) Human U1 small nuclear RNA genes: extensive conservation of flanking sequences suggests cycles of gene amplification and transposition. Mol Cell Biol 5:2159–2171

Bindereif A, Wolff T, Green MR (1990) Discrete domains of human U6 snRNA required for the assembly of U4/U6 snRNP and splicing complexes. EMBO J 9:251–255

Birnstiel ML, Schaufele FJ (1988) Structure and function of minor U snRNPs. In: Birnstiel M (ed) Structure and function of major and minor snRNPs. Springer, Berlin Heidelberg New York Tokyo, pp 155–182

Black DL, Steitz JA (1986) Pre-mRNA splicing in vitro requires intact U4/U6 small nuclear ribonucleoprotein. Cell 46:697–704

Blatt C, Saxe D, Marzluff WF, Lobo S, Nesbitt MN, Simon MI (1988) Mapping and gene order of U1 small nuclear RNA, endogenous viral env sequence, amylase and alcohol dehydrogenase-3 on mouse chromosome 3. Somatic Cell Mol Genet 14:133–142

Bogenhagen DF, Brown DD (1981) Nucleotide sequence in *Xenopus* 5S DNA required for transcription termination. Cell 24:261–270

Bohmann D, Keller W, Dale T, Scholer HR, Tebb G, Mattaj IW (1987) A transcription factor which binds to enhancers of SV40, immunoglobulin heavy chain and U2 snRNA genes. Nature (Lond) 325:268–272

Brennwald P, Porter G, Wise JA (1988) U2 small nuclear RNA is remarkably conserved between *Schizosaccharomyces pombe* and mammals. Mol Cell Biol 8:5575–5580

Bringmann P, Luhrmann R (1986) Purification of individual snRNPs U1, U2, U5 and U4/U6 from HeLa cells and characterization of their protein components. EMBO J 5:3509–3516

Brow DA, Guthrie C (1988) Spliceosomal RNA U6 is remarkably conserved from yeast to mammals. Nature (Lond) 334:213–218

Brow DA, Guthrie C (1990) Transcription of a yeast U6 snRNA gene requires a pol III promoter element in a novel position. Genes Develop 4:1345–1356

Brown DT, Morris GF, Chodchoy N, Sprecher C, Marzluff WF (1985) Structure of the sea urchin U1 RNA repeat. Nucl Acids Res 13:537–555

Brown JWS, Waugh R (1989) Maize U2 snRNA gene sequence and expression. Nucl Acids Res 17:8991–9001

Brunel C, Sri-Widada J, Jeanteur P (1985) snRNPs and scRNPs in eukaryotic cells. Prog Mol Subcell Biol 9:1–52

Bruznik JP, Doren KV, Hirsh D, Steitz JA (1988) *Trans*-splicing involves a novel form of small nuclear ribonucleoprotein particles. Nature (Lond) 335:559–562

Buckland RA, Cooke HJ, Roy KL, Dahlberg JE, Lund E (1983) Isolation and characterization of three cloned fragments of human DNA coding for tRNAs and small nuclear RNA U1. Gene 22:211–217

Busch H, Reddy R, Rothblum L, Choi YC (1982) SnRNAs, snRNPs, and RNA processing. Ann Rev Biochem 51:617–654

Carbon P, Murgo S, Ebel J-P, Krol A, Tebb G, Mattaj IW (1987) A common octamer motif binding protein is involved in the transcription of U6 snRNA by RNA polymerase III and U2 snRNA by RNA polymerase II. Cell 51:71–79

Card CO, Morris GF, Brown DT, Marzluff WF (1982) Sea urchin small nuclear RNA genes are organized in distinct tandemly repeating units. Nucl Acids Res 10:7677–7688

Chabot B, Black DL, Steitz JA (1985) The 3′ splice site of pre-messenger RNA is recognized by a small nuclear ribonucleoprotein. Science 230:1344–1349

Chandrasekharappa SC, Smith JH, Eliceiri GL (1983) Biosynthesis of small nuclear RNAs in human cells. J Cell Physiol 117:169–174

Christofori G, Keller W (1988) 3′ cleavage and polyadenylation of mRNA precursors in vitro requires a poly(A) polymerase, a cleavage factor, and a snRNP. Cell 54:875–889

Ciliberto G, Buckland R, Cortese R, Philipson L (1985) Transcription signals in embryonic *Xenopus laevis* U1 RNA genes. EMBO J 4:1537–1543

Corden J, Wasylyk B, Buchwalder A, Sassone-Corsi P, Kedinger C, Chambon P (1980) Promoter sequences of eukaryotic protein-coding genes. Science 209:1406–1414

Dahlberg JE, Lund E (1988) The genes and transcription of major small nuclear RNAs. In: Birnstiel M (ed) Structure and function of major and minor snRNPs. Springer, Berlin Heidelberg New York Tokyo, pp 38–70

Dahlberg JE, Schenborn ET (1988) The human U1 snRNA promoter and enhancer do not direct synthesis of messenger RNA. Nucl Acids Res 16:5827–5840

Das G, Henning D, Reddy R (1987) Structure, organization and transcription of *Drosophila* U6 small nuclear RNA genes. J Biol Chem 262:1187–1193

Das G, Henning D, Wright D, Reddy R (1988) Upstream regulatory elements are necessary and sufficient for transcription of a U6 RNA gene by RNA polymerase III. EMBO J 7:503–512

Denison RA, Van Arsdell SW, Bernstein LB, Weiner AM (1981) Abundant pseudogenes for small nuclear RNAs are dispersed in the human genome. Proc Natl Acad Sci USA 78:810–814

Earley JM, Roebuck KA, Stumph WE (1984) Three linked chicken U1 RNA genes have limited flanking DNA sequence homologies that reveal potential regulatory signals. Nucl Acids Res 12:7411–7421

Edery I, Sonenberg N (1985) Cap-dependent RNA splicing in a HeLa nuclear extract. Proc Natl Acad Sci USA 82:7590–7594

Eliceiri GL (1980) Formation of low molecular weight RNA species in HeLa cells. J Cell Physiol 102:199–207

Eliceiri GL, Sayaverdra MS (1976) Small RNAs is the nucleus and cytoplasm of HeLa cells. Biochem Biophys Res Commun 72:507–512

Eliceiri GL, Gurney T Jr (1978) Subcellular location of precursors to small nuclear RNA species C and D and of newly synthesized 5S RNA in HeLa cells. Biochem Biophys Res Commun 81:915–919

Eliceiri GL, Smith JH (1983) Sensitivity to UV radiation of small nuclear RNA synthesis in mammalian cells. Mol Cell Biol 3:2151–2155

Epstein P, Reddy R, Henning D, Busch H (1980) The nucleotide sequence of nuclear U6 (4.7S) RNA. J Biol Chem 255:8901–8906

Fisher U, Luhrmann R (1990) The m_3G cap is essential for the nuclear transport of U snRNPs and may function as part of a snRNP-specific nuclear location signal. Science 249:786–790

Flanegan JB, Pettersson RF, Ambros V, Hewlett MJ, Baltimore D (1977) Covalent linkage of a protein to a defined nucleotide sequence at the 5'-terminus of a virion and replicative intermediate RNAs of poliovirus. Proc Natl Acad Sci USA 74:961–965

Forbes DJ, Kornberg T, Kirschner MW (1983) Small nuclear RNA transcription and ribonucleo protein assembly in early *Xenopus* development. J Cell Biol 97:62–72

Forbes DJ, Kirschner MW, Caput D, Dahlberg JE, Lund E (1984) Differential expression of multiple U1 small nuclear RNAs in oocytes and embryos of *Xenopus laevis*. Cell 38:681–689

Frederiksen S, Hellung-Larsen P, Gram-Jensen E (1978) The differential inhibitory effect of α-amanitin on the synthesis of low molecular weight components in BHK cells. FEBS Lett 87:227–231

Frendeway D, Barta I, Gillespie M, Potashkin J (1990) Schizosaccharomyces U6 genes have a sequence within their introns that matches the B box consensus of tRNA internal promoters. Nucl Acids Res 18:2025–2032

Fritz A, Parisot RF, Newmeyer D, De Robertis EM (1984) Small nuclear U RNPs in *Xenopus laevis* development: uncoupled accumulation of the protein and RNA components. J Mol Biol 178:273–285

Fuerst TR, Moss B (1989) Structure and stability of mRNA synthesized by vaccinia virus-encoded bacteriophage T7 RNA polymerase in mammalian cells J Mol Biol 206:333–348

Furuichi Y (1978) Pretranscriptional capping in the biosynthesis of cytoplasmic polyhedrosis virus mRNA. Proc Natl Acad Sci USA 82:488–492

Geiduschek EP, Tocchini-Valentini GP (1988) Transcription by RNA polymerase III. Ann Rev Biochem 57:873–914

Georgiev O, Mous J, Birnstiel M (1984) Processing and nucleo-cytoplasmic transport of histone gene transcripts. Nucl Acids Res 12:8539–8551

Gottlieb E, Steitz JA (1989) Function of the mammalian La protein: evidence for its action in transcription termination by RNA polymerase III. EMBO J 8:851–861

Gram-Jensen E, Hellung-Larsen P, Frederiksen S (1979) Synthesis of low molecular weight components A, C and D by polymerase II in α-amanitin-resistant hamster cells. Nucl Acids Res 6:321–330

Green M (1986) Pre-mRNA splicing. Ann Rev Genet 20:671–708

Gunderson SI, Murphy JT, Knuth MW, Steinberg TH, Dahlberg JE, Burgess RR (1988) Binding of transcription factors to the promoter of the human U1 RNA gene studied by foot-printing. J Biol Chem 263:17603–17610

Gupta S, Singh R, Reddy R (1990a) Capping of U6 small nuclear RNA in vitro can be uncoupled from transcription. J Biol Chem 565:9491–9495

Gupta S, Busch R, Singh R, Reddy R (1990b) Characterization of U6 snRNA cap-specific antibodies: Identification of γ-monomethy GTP cap structure in 7SK and several other human small RNAs. J Biol Chem (in press)

Guthrie C, Patterson B (1988) Spliceosomal snRNAs. Ann Rev Gen 22:387–419

Hamada K, Kumazaki T, Mizuno K, Yokoro K (1989) A small nuclear RNA, U5, can transform cells in vitro. Mol Cell Biol 9:4345–4356

Hamm J, Mattaj IW (1989) An abundant U6 snRNP found in germ cells and embryos of *Xenopus laevis*. EMBO J 8:4179–4187

Hamm J, Darzynkiewicz E, Tahara SM, Mattaj IW (1990) The trimethylguanosine cap structure of U1 snRNA is a component of a bipartite nuclear targeting signal. Cell 62:569–577

Hanley BA, Schuler MA (1989) Nucleotide sequence of a pea U2 snRNA gene. Nucl Acids Res 17:10106

Hart RP, McDevitt MA, Nevins JR (1985) Poly (A) site cleavage in a HeLa nuclear extract is dependent on downstream sequences. Cell 43:677–683

Hashimoto C, Steitz JA (1983) Sequential association of 7–2 RNA with two different autoantigens. J Biol Chem 258:1379–1382

Hellung-Larsen P, Kulamowicz I, Frederiksen S (1980) Synthesis of low molecular weight RNA components in cells with a temperature-sensitive polymerase II. Biochem Biophys Acta 609:201–204

Hellung-Larsen P, Gram-Jensen E, Frederiksen S (1981) Effect of 5,6-Dichloro-1-b-D-ribofuranosyl benzimidazole on the synthesis of low molecular weight RNA components. Biochem Biophys Res Commun 99:1303–1310

Hernandez N (1985) Formation of the 3′ end of U1 snRNA is directed by a conserved sequence located downstream of the coding region. EMBO J 4:1827–1837

Hernandez N, Weiner AM (1986) Formation of the 3′ end of U1 snRNA requires compatible snRNA promoter elements. Cell 47:249–258

Hernandez N, Lucito R (1988) Elements required for transcription initiation of the human U2 snRNA gene coincide with elements required for snRNA 3′ end formation. EMBO J 7:3125–3134

Hodnett JL, Busch H (1968) Isolation and characterization of uridylic acid-rich 7S ribonucleic acid of rat liver nuclei. J Biol Chem 243:6334–6342

Hoffman ML, Korf GM, McNamara KJ, Stumph WE (1986) Structural and functional analysis of chicken U4 small nuclear RNA genes. Mol Cell Biol 6:3910–3919

Horikoshi M, Wang CK, Fujii H, Cromlish JA, Weil PA, Roeder RG (1989) Cloning and structure of a yeast gene encoding a general transcription initiation factor TFIID that binds to the TATA box. Nature (Lond) 341:299–303

Howard EF, Michael SK, Dahlberg JE, Lund E (1986) Functional, developmentally expressed genes for mouse U1a and U1b snRNAs contain both conserved and non-conserved transcription signals. Nucl Acids Res 14:9811–9824

Hughes JMX, Konings DAM, Cesareni G (1987) The yeast homologue of U3 snRNA. EMBO J 6:2145–2155

Janson L, Pettersson U (1990) Cooperative interactions between transcription factors Sp1 and OTF-1. Proc Natl Acad Sci USA 87:4732–4736

Jelinek WR, Schmidt CW (1982) Repetitive sequences in eukaryotic DNA and their expression. Ann Rev Biochem 51:813–844

Kass S, Tyc K, Steitz JA, Sollner-Webb B (1990) The U3 small nucleolar ribonucleoprotein functions in the first step of preribosomal RNA processing. Cell 60:897–908

Kassavetis GA, Braun BR, Nguyen LH, Geiduschek EP (1990) S. cerevisiae TFIIIB is the transcription initiation factor proper of RNA polymerase III, while TFIIIA and TFIIIC are assembly factors. Cell 60:235–245

Kato N, Harada F (1985) New U1 RNA species found in friend SFFV (Spleen Focus Forming virus)-transformed cells. J Biol Chem 260:7775–7782

Kazmaier M, Tebb G, Mattaj IW (1987) Functional characterization of X. laevis U5 snRNA genes. EMBO J 6:3071–3078

Kejzlarova-Lepesant J, Brock HW, Moreau J, Dubertret ML, Billault A, Lepasant JA (1984) A complete and truncated U1 snRNA gene of Drosophilla melanogaster are found as inverted repeats at region 82E of the polytene chromosomes. Nucl Acids Res 12:8835–8846

Kiss T, Solymosy F (1990) Molecular analysis of a U3 RNA gene locus in tomato: transcription signals, the coding region, expression in transgenic tobacco plants and tandemly repeated pseudogenes. Nucl Acids Res 18:1941–1949

Kleinert H, Sebastian B, Benecke J (1990) Expression of a human 7SK RNA gene in vivo requires a novel pol III upstream element. EMBO J 9:711–718

Kleinschmidt AM, Pederson T (1987) Accurate and efficient 3′ processing of U2 small nuclear RNA precursor in a fractionated extract. Mol Cell Biol 7:3131–3137

Kleinschmidt AM, Pederson T (1990) RNA processing and ribonucleoprotein assembly studied in vivo by RNA transfection. Proc Natl Acad Sci USA 87:1283–1287

Kleinschmidt AM, Pederson T, Tani T, Ohshima Y (1990) An intron-containing Schizosaccharomyces pombe U6 RNA gene can be transcribed by human RNA polymerase III. J Mol Biol 211:7–9

Knuth MW, Gunderson SI, Thompson NE, Strasheim LA, Burgess RR (1990) Purification and characterization of PSE1, a transcription activating protein related to Ku and TREF that binds the proximal sequence element of the human U1 promoter. J Biol Chem 265:17911–17920

Konarska MM, Padgett RA, Sharp PA (1984) Recognition of cap structure in splicing in vitro of mRNA precursors. Cell 38:731–736

Korf GM, Stumph WE (1986) Chicken U2 and U1 RNA genes are found in very different genomic environments but have similar promoter structures. Biochemistry 25:2041–2047

Korf GM, Stumph WE (1989) Multiple functional motifs in the enhancers of chicken U1 and U4 small nuclear RNA genes. UCLA Symp 94:165–175

Korf GM, Botros IW, Stumph WE (1988) Developmental and tissue-specific expression of U4 small nuclear RNA genes. Mol Cell Biol 8:5566–5569

Kretzner L, Rymond BC, Rosbash M (1987) *S. cerevisiae* U1 RNA is large and has limited primary sequence homology to metazoan U1 snRNA. Cell 50:593–602

Krol A, Ebel J-P, Rinke J, Luhrmann R (1983) U1, U2, and U5 small nuclear RNAs are found in plant cells. Complete nucleotide sequence of the U5 RNA family from pea nuclei. Nucl Acids Res 11:8583–8594

Krol A, Lund E, Dahlberg JE (1985) The two embryonic U1 RNA genes of *Xenopus laevis* have both common and gene-specific transcription signals. EMBO J 4:1529–1535

Krol A, Carbon P, Ebel J-P, Appel B (1987) *Xenopus tropicalis* U6 snRNA genes transcribed by Pol III contain the upstream promoter elements used by pol II-dependent U snRNA genes. Nucl Acid Res 15:2463–2478

Kruger W, Benecke B-J (1987) Structural and functional analysis of a human 7SK RNA gene. J Mol Biol 195:31–41

Kunkel GR, Pederson T (1988) Upstream elements required for efficient transcription of a human U6 RNA gene resemble those of U1 and U2 genes even though a different polymerase is used. Genes Develop 2:196–204

Kunkel GR, Pederson T (1989) Transcription of a human U6 small nuclear RNA gene in vivo withstands deletion of intragenic sequences but not of an upstream TATATA box. Nucl Acids Res 17:7371–7379

Kunkel GR, Maser RI, Calvet JP, Pederson T (1986) U6 small nuclear RNA is transcribed by RNA polymerase III. Proc Natl Acad Sci USA 83:8575–8579

Lee SI, Murthy SCS, Trimble JJ, Desrosiers RC, Steitz JA (1988) Four novel U RNAs are encoded by a herpesvirus. Cell 54:599–607

Lerner M, Steitz JA (1979) Antibodies to small nuclear RNAs complexed with proteins are produced by patients with systemic lupus erythematosus. Proc Natl Acad Sci USA 76:5495–5499

Li HV, Zagorski J, Fournier MJ (1990) Depletion of U14 small nuclear RNA (snR128) disrupts production of 18S rRNA in *Saccharomyces cerevisiae*. Mol Cell Biol 10:1145–1152

Lindgren V, Ares M Jr, Weiner AM, Francke U (1985a) Human genes for U2 small nuclear RNA map to a major adenovirus 12 modification site on chromosome 17. Nature (Lond) 314:115–116

Lindgren V, Bernstein LB, Weiner AM, Francke U (1985b) Human U1 small nuclear RNA pseudogenes do not map to the site of the U1 genes in 1p36 but are clustered in 1q12–q22. Mol Cell Biol 5:2172–2180

Lobo SM, Marzluff WF (1987) Synthesis of U1 RNA in isolated mouse cell nuclei: initiation and 3′-end formation. Mol Cell Biol 7:4290–4296

Lobo SM, Hernandez N (1989) A 7 bp mutation converts a human RNA polymerase II snRNA promoter into an RNA polymerase III promoter. Cell 58:55–67

Lobo SM, Ifill S, Hernandez N (1990) *Cis*-acting elements required for RNA polymerase II and III transcription in the human U2 and U6 snRNA promoters. Nucl Acids Res 18:2891–2899

Long EO, Dawid IB (1980) Repeated genes in eukaryotes. Ann Rev Biochem 49:727–764

Lorenzi MD, Rohrer U, Birnsteil ML (1986) Analysis of a sea urchin gene cluster coding for the small nuclear U7 RNA, a rare RNA species implicated in the 3′ editing of histone precursor mRNAs. Proc Natl Acad Sci USA 83:3243–3247

Lund E (1988) Heterogeneity in human U1 snRNAs. Nucl Acids Res 16:5813–5826

Lund E, Dahlberg JE (1984) True genes for human U1 small nuclear RNA: copy number, polymorphism and methylation. J Biol Chem 259:2013–2021

Lund E, Dahlberg JE (1987) Differential accumulation of U1 and U4 small nuclear RNAs during *Xenopus* development. Genes Develop 1:39–46

Lund E, Dahlberg J (1989) In vitro synthesis of vertebrate U1 snRNA. EMBO J 8:287–292

Lund E, Nesbitt MN (1988) Embryonic and adult mouse U1 snRNA genes map to different chromosomal loci. Somatic Cell Mol Genet 14:143–148

Lund E, Bostock CJ, Robertson M, Christie S, Mitchen JL, Dahlberg JE (1983) U1 small nuclear RNA genes are located on human chromosome 1 and are expressed in mouse-human hybrids. Mol Cell Biol 3:2211–2220

Lund E, Dahlberg JE, Forbes DJ (1984) The two embryonic U1 small nuclear RNAs of *Xenopus laevis* are encoded by a major family of tandemly repeated genes. Mol Cell Biol 4:2580–2586

Lund E, Kahan B, Dahlberg JE (1985) Differential control of U1 small nuclear RNA expression during mouse development. Science 229:1271–1274

Lund E, Bostock CJ, Dahlberg JE (1987) The transcription of *Xenopus laevis* embryonic U1 snRNA genes changes when oocytes mature into eggs. Genes Develop 1:47–56

Madore SJ, Wieben ED, Pederson T (1984a) Intracellular site of U1 small nuclear RNA processing and ribonucleoprotein assembly. J Cell Biol 98:188–192

Madore SJ, Wieben ED, Kunkel GR, Pederson T (1984b) Precursors of U4 small nuclear RNA. J Cell Biol 99:1140–1144

Mangin M, Ares M Jr, Weiner AM (1985) U1 small nuclear RNA genes are subject to dosage compensation. Science 229:272–275

Mangin M, Ares M Jr, Weiner AM (1986) Human U2 small nuclear RNA genes contain an upstream enhancer. EMBO J 5:987–995

Maniatis T, Reed R (1987) The role of small nuclear ribonucleoprotein particles in pre-mRNA splicing. Nature (Lond) 325:673–678

Manser T, Gesteland RF (1981) Characterization of small nuclear RNA U1 gene candidates and pseudogenes from the human genome. J Mol Appl Genet 1:117–125

Manser T, Gesteland RF (1982) Human U1 loci: genes for human U1 RNA have dramatically similar genomic environment. Cell 29:257–264

Marzluff WF, Brown DT, Lobo S, Wang S (1983) Isolation and characterization of two linked mouse U1b small nuclear RNA genes. Nucl Acids Res 11:6255–6270

Maser RL, Calvet JP (1989) U3 small nuclear RNA can be cross-linked in vivo to the 5′ external transcribed spacer of pre-ribosomal RNA. Proc Natl Acad Sci USA 86:6523–6527

Mattaj IW (1986) Cap trimethylation of U snRNA is cytoplasmic and dependent on U snRNA protein binding. Cell 46:905–911

Mattaj IW, Zeller R (1983) *Xenopus laevis* U2 snRNA genes: tandemly repeated transcription units sharing 5′ and 3′ flanking homology with other RNA polymerase II transcribed genes. EMBO J 2:1883–1891

Mattaj IW, Leinhard S, Jiricny J, De Robertis EM (1985) An enhancer-like sequence within the *Xenopus* U2 gene promoter facilitates the formation of stable transcription complexes. Nature (Lond) 316:163–167

Mattaj IW, Dathan NA, Parry HD, Carbon P, Krol A (1988) Changing the RNA polymerase specificity of U snRNA gene promoters. Cell 55:435–442

Mazan S, Bachellerie J-P (1988) Structure and organization of mouse U3B functional genes. J Biol Chem 263:19461–19467

McNamara KJ, Stumph W (1990) Site-directed mutational analysis of a U4 small nuclear RNA gene proximal sequence element. J Biol Chem 265:9728–9731

McNamara KJ, Walker RJ, Roebuck KA, Stumph WE (1987) Transcriptional signals of a U4 small nuclear RNA gene. Nucl Acids Res 15:9239–9254

Michael SK, Hilgers J, Kozak C, Whitney JB, Howard F (1986) Characterization and mapping of DNA sequence homologous to mouse U1a snRNA: localization on chromosome 11 near the Dlb-1 and Re loci. Somatic Cell Mol Gen 12:215–223

Moenne A, Camier S, Anderson G, Morgottin F, Beggs J, Sentenac A (1990) The U6 gene of *Saccharomyces cerevisiae* is transcribed by RNA polymerase C (III) in vivo and in vitro. EMBO J 9:271–277

Monstein H-J, Hammarstrom K, Westin G, Zabielski J, Philipson L, Pettersson U (1983) Loci for human U1 RNA: structural and evolutionary implications. J Mol Biol 167:245–257

Montzka KA, Steitz JA (1988) Additional low-abundance human small nuclear ribonucleoproteins: U11, U12, etc. Proc Natl Acad Sci USA 85:8885–8889

Morra DS, Lawler SH, Eliceiri BP, Eliceiri GL (1986) Inhibition of small nuclear RNA synthesis by ultraviolet radiation. J Biol Chem 261:3142–3146

Morris GF, Marzluff W (1985) Synthesis of U1 RNA in isolated nuclei from sea urchin embryos: U1 RNA is initiated at the first nucleotide of the RNA. Mol Cell Biol 5:1143–1150

Morris GF, Price DH, Marzluff WF (1986) Synthesis of U1 RNA in a DNA-dependent system from sea urchin embryos. Proc Natl Acad Sci USA 83:3674–3678

Moshier JA, Deutch AH, Huang RCC (1987) Structure and in vitro transcription of a mouse B1 cluster containing a unique B1 dimer. Gene 58:19–27

Moshier JA, Deutch AH, Huang RCC (1988) Sequence of a mouse U2 snRNA gene expressed in transfected mouse cells. Nucl Acids Res 16:7203

Mottram J, Perry KL, Lizardi PM, Luhrmann R, Agabian N, Nelson RG (1989) Isolation and sequence of four small nuclear U RNA genes of *Trypanosoma brucei*: subsp. *brucei*. Identification of the U2, U4 and U6 RNA analogs. Mol Cell Biol 9:1212–1223

Mount SM, Steitz JA (1981) Sequence of U1 RNA from *Drosophila melanogaster*: implications for U1 secondary structure and possible involvement in splicing. Nucl Acids Res 9:6351–6368

Mount SM, Pettersson I, Hinterberger M, Karmas A, Steitz JA (1981) The U1 small nuclear RNA, protein complex selectively binds a 5' splice site in vitro. Cell 33:509–518

Moussa NM, E-L-Din AS, Lobo SM, Marzluff WF (1987) A mouse U1b-2 gene with extensive sequence similarity to a rat U1a gene for 670 nucleotides 5' to the gene. Nucl Acids Res 15:3622

Mowry KL, Steitz JA (1987) Identification of the human U7 snRNP as one of several factors involved in the 3' end maturation of histone pre-mRNAs. Science 238:1682–1687

Murphy JT, Burgess RR, Dahlberg JE, Lund E (1982) Transcription of a gene for human U1 small nuclear RNA. Cell 29:265–274

Murphy S, Tripodi M, Melli M (1986) A sequence upstream from the coding region is required for the transcription of the 7SK RNA genes. Nucl Acids Res 14:9243–9260

Murphy JT, Skuzeski JM, Lund E, Steinberg TH, Burgess RR, Dahlberg JE (1987a) Functional elements of the human U1 RNA promoter. J Biol Chem 262:1795–1803

Murphy S, DiLiegro C, Melli M (1987b) The in vitro transcription of the 7SK RNA gene by RNA polymerase III is dependent only on the presence of an upstream promoter. Cell 51:81–87

Murphy S, Pierani A, Scheidereit C, Melli M, Roeder R (1989) Purified octamer binding transcription factors stimulate RNA polymerase III-mediated transcription of the 7SK RNA gene. Cell 59:1071–1080

Myslinski E, Segault V, Branlant C (1990) An intron in the genes for U3 small nucleolar RNAs of the yeast *Saccharomyces cerevisiae*. Science 247:1213–1216

Nash MA, Marzluff WF (1988) Structure of an unusual sea urchin U1 RNA gene cluster. Gene 64:53–63

Nash MA, Sakallah S, Santiago C, Yu J-C, Marzluff WF (1989) A developmental switch in sea urchin U1 RNA. Dev Biol 134:289–296

Naylor SL, Zabel BU, Manser T, Gesteland R, Sakaguchi AY (1984) Localization of human U1 small nuclear RNA genes to band p36.3 of chromosome 1 by in situ hybridization. Somatic Cell Mol Genet 10:307–313

Neuman de Vegvar HE, Dahlberg JE (1989) Initiation and termination of human U1 RNA transcription requires the concerted action of multiple flanking elements. Nucl Acids Res 17:9305–9318

Neuman de Vegvar HE, Dahlberg JE (1990) Nucleocytoplasmic transport and processing of small nuclear RNA precursors. Mol Cell Biol 10:3365–3375

Neuman de Vegvar HE, Lund E, Dahlberg JE (1986) 3' end formation of U1 snRNA precursors is coupled to transcription from snRNA promoters. Cell 47:259–266

Nojima H, Kornberg RD (1983) Genes and pseudogenes for mouse U1 and U2 small nuclear RNAs. J Biol Chem 258:8151–8155

Ohshima Y, Okada N, Tani T, Itoh Y, Itoh M (1981) Nucleotide sequences of mouse genomic loci including a gene or pseudogene for U6 (4.8S) nuclear RNA. Nucl Acids Res 9:5145–5158

Padgett RA, Grabowski PJ, Konarska MM, Seiler S, Sharp PA (1986) Splicing of messenger RNA precursors. Ann Rev Biochem 55:1119–1150

Parker R, Siliciano PG, Guthrie C (1987) Recognition of the TACTAAC box during mRNA splicing in yeast involves base-pairing to the U2-like snRNA. Cell 49:229–239

Parker R, Simmons T, Shuster EO, Siliciano PG, Guthrie C (1988) Genetic analysis of small nuclear RNAs in *Saccharomyces cerevisiae*: viable sextuple mutant. Mol Cell Biol 8:3150–3159

Parry HD, Mattaj IW (1990) Positive and negative functional interactions between promoter elements from different classes of RNA polymerase III-transcribed genes. EMBO J 9:1097–1104

Parry HD, Tebb G, Mattaj IW (1989) The *Xenopus* gene PSE is single, compact element required for transcription initiation and 3' end formation. Nucl Acids Res 17:3633–3649

Patterson B, Guthrie C (1987) An essential yeast snRNA with a U5-like domain is required for splicing in vivo. Cell 49:613–624

Paule MR (1990) In search of a single factor. Nature (Lond) 344:819–820

Porter GL, Brennwald PJ, Holm KA, Wise JA (1988) The sequence of U3 from *Schizosaccharomyces pombe* suggests structural divergence of this snRNA between metazoans and unicellular eukaryotes. Nucl Acids Res 16:10131–10152

Reddy R (1988) Transcription of a U6 small nuclear RNA gene in vitro. J Biol Chem 263:15980–15984

Reddy R, Busch H (1988) Small nuclear RNAs: RNA sequences, structure, and modifications. In: Birnstiel M (ed) Structure and function of major and minor snRNPs. Springer, Berlin Heidelberg New York Tokyo, pp 1–37

Reddy R, Henning D, Busch H (1979) Nucleotide sequence of nucleolar U3B RNA. J Biol Chem 254:220–224

Reddy R, Henning D, Busch H (1985) Primary and secondary structure of U8 small nuclear RNA. J Biol Chem 260:10930–10935

Reddy R, Henning D, Das G, Harless M, Wright D (1987) The capped U6 small nuclear RNA is transcribed by RNA polymerase III. J Biol Chem 262:75–81

Reich C, Wise JA (1990) Evolutionary origin of the U6 snRNA intron. Mol Cell Biol

Riedel N, Wise JA, Swerdlow H, Mak A, Guthrie C (1986) Small nuclear RNAs from *Saccharomyces cerevisiae*: unexpected diversity in abundance, size, and molecular complexity. Proc Natl Acad Sci USA 83:8097–8101

Rinke J, Steitz JA (1985) Association of the lupus antigen La with subset of U6 snRNA molecules. Nucl Acids Res 13:2617–2629

Roberts MR, Miskimins WK, Ruddle FH (1989) Nuclear proteins TREF1 and TREF2 bind to the transcriptional control element of the transferrin receptor gene and appear to be associated as a heterodimer. Cell Regulation 1:151–164

Ro-Choi TS, Raj NBK, Pike LM, Busch H (1976) Effects of α-amanitin, cycloheximide, and thioacetamide on low molecular weight nuclear RNA. Biochemistry 15:3823–3828

Roebuck KA, Walker RJ, Stumph WE (1987) Multiple functional motifs in the chicken U1 RNA gene enhancer. Mol Cell Biol 7:4185–4193

Roebuck KA, Szeto DP, Green KP, Fan QN, Stumph WE (1990) Octamer and SPH motifs in the U1 enhancer cooperate to activate U1 RNA gene expression. Mol Cell Biol 10:341–352

Roiha H, Shuster EO, Brow DA, Guthrie C (1989) Small nuclear RNAs from budding yeasts: phylogenetic comparisons reveal extensive size variation. Gene 82:137–144

Roop DR, Kristo P, Stumph WE, Tsai MJ, O'Malley BW (1981) Structure and expression of a chicken gene coding for U1 RNA. Cell 23:671–680

Saba JA, Busch H, Wright D, Reddy R (1986) Isolation and characterization of two putative full-length *Drosophila* U4 small nuclear RNA genes. J Biol Chem 261:8750–8753

Saluz HP, Schmidt T, Dudler R, Atlwegg M, Stumm-Zollinger E, Kubli E, Chen PS (1983) The genes coding for 4 snRNAs of *Drosophila melanogaster* localization and determination of gene numbers. Nucl Acids Res 11:77–90

Saluz HP, Dudler R, Schmidt T, Kubli E (1988) The localization and estimated copy number of *Drosophila melanogaster* U1, U4, U5 and U6 snRNA genes. Nucl Acids Res 16:3582

Schaufele F, Gilmartin GM, Bannwarth W, Birnstiel ML (1986) Compensatory mutations suggest that base-pairing with a small nuclear RNA is required to form the 3' end of H3 messenger RNA. Nature (Lond) 323:777–781

Shatkin AJ (1976) Capping of eukaryotic mRNAs. Cell 9:645–653

Shatkin AJ (1985) mRNA cap binding proteins: essential factors for initiating translation. Cell 40:223–224

Siliciano PG, Jones MH, Guthrie C (1987a) *Saccharomyces cerevisiae* has a U1-like small nuclear RNA with unexpected properties. Science 237:1484–1487

Siliciano PG, Brow DA, Roiha H, Guthrie C (1987b) An essential snRNA from *S. cerevisiae* has properties predicted for U4, including interaction with a U6-like snRNA. Cell 50:585–592

Simpson L, Shaw J (1989) RNA editing and the mitochondrial cryptogenes of kinetoplastid protozoa. Cell 57:355–366

Singh R, Reddy R (1989) γ-Monomethyl phosphate: A cap structure in spliceosomal U6 small nuclear RNA. Proc Natl Acad Sci US 86:8280–8283

Singh R, Gupta S, Reddy R (1990) Capping of mammalian U6 small nuclear RNA in vitro is directed by a conserved stem-loop and AUAUAC sequence: conversion of a noncapped RNA into a capped RNA. Mol Cel Biol 10:939–946

Skinner HB, Adams DS (1987) Nucleotide sequence of *Physarum* U6 small RNA. Nucl Acids Res 15:371

Skuzeski JM, Lund E, Murphy JT, Steinberg TH, Burgess RR, Dahlberg JE (1984) Synthesis of human U1 RNA: identification of two regions of the promoter essential for transcription initiation at position +1. J Biol Chem 259:8345–8352

Southgate C, Busslinger M (1989) In vivo and in vitro expression of U7 snRNA genes: *cis-* and *trans*-acting elements required for RNA polymerase II-directed transcription. EMBO J 8:539–549

Steinberg TH, Mathews DE, Durbin RD, Burgess RR (1990) Tagetitoxin: a new inhibitor of eukaryotic transcription by RNA polymerase III. 265:499–505

Steitz JA (1988) "SNURPS". Sci Am 258:56–61

Steitz JA, Black D, Gerke V, Parker KA, Kramer A, Frendeway D, Keller W (1988) Function of abundant U snRNPs. In: Birnstiel M (ed) Structure and function of major and minor snRNPs. Springer, Berlin Heidelberg New York Tokyo, pp 115–154

Stroke IL, Weiner AM (1985) Genes and pseudogenes for rat U3A and U3B small nuclear RNA. J Mol Biol 184:183–193

Stroke I, Weiner A (1989) The 5' end of U3 snRNA can be cross-linked in vivo to the external transcribed spacer of rat ribosomal RNA precursors. J Mol Biol 210:497–512

Strub K, Birnstiel M (1986) Genetic complementation in the *Xenopus* oocyte: co-expression of sea urchin histone and U7 RNAs restores 3' processing of H3 pre-mRNA in the oocyte. EMBO J 5:1675–1682

Suh D, Busch H, Reddy R (1986) Isolation and characterization of a human U3 small nucleolar RNA gene. Biochem Biophys Res Commun 137:1133–1140

Szkukalek, Kiss T, Solymosy F (1990) The 5' end of the coding region of a U6 RNA gene candidate from tomato starts with GUCC, a phylogenetically highly conserved 5' end sequence of U6 RNA. Nucl Acids Res 18:1295

Tan E (1989) Antinuclear antibodies: diagnostic markers for autoimmune diseases and probes for cell biology. Adv Immunol 44:93–151

Tanaka M, Grossniklaus U, Herr W, Hernandez N (1988) Activation of the U2 snRNA promoter by the octamer motif defines a new class of RNA polymerase II enhancer elements. Genes Develop 2:1764–1778

Tani T, Ohshima Y (1989) The gene for U6 small nuclear RNA in fission yeast has an intron. Nature (Lond) 337:87–90

Tani T, Watanabe-Nagasu N, Okada N, Ohshima Y (1983) Molecular cloning and characterization of a gene for rat U2 small nuclear RNA. J Mol Biol 168:579–594

Tazi J, Alibert C, Temsamani J, Reveillaud I, Cathala G, Brunel C, Jeanteur P (1986) A protein that specifically recognizes the 3' splice site of mammalian pre-mRNA introns is associated with a small nuclear ribonucleoprotein. Cell 47:755–766

Tebb G, Mattaj IW (1989) The *Xenopus laevis* U2 gene distal sequence element (enhancer) is composed of four subdomains that can act independently and are partly functionally redundant. Mol Cell Biol 9:1682–1690

Tebb G, Bohmann D, Mattaj IW (1987) Only two of the four sites of interaction with nuclear factors within the *Xenopus* U2 gene promoter are necessary for efficient transcription. Nucl Acids Res 15:6437–6453

Thomas JD, Conrad RC, Blumenthal T (1988) The *C. elegans trans*-spliced leader RNA is bound to Sm and has a trimethylguanosine cap. Cell 53:533–539

Thomas JD, Lea K, Zucker-Aprison E, Blumenthal T (1990) *C. elegans* snRNAs. Nucl Acids Res 18:2633–2642

Thompson NE, Steinberg TH, Aronson DB, Burgess R (1989) Inhibition of in vivo and in vitro transcription by monoclonal antibodies prepared against wheat germ RNA polymerase II that react with heptapeptide repeat of eukaryotic polymerase II. J Biol Chem 264:11511–11520

Tollervey D, Wise JA, Guthrie C (1983) A U4-like small nuclear RNA is dispensable in yeast. Cell 35:753–762

Tsuchidi C, Ullu E (1990) Destruction of U2, U4, or U6 small nuclear RNA blocks *trans*-splicing in *Trypanosoma* cells. Cell 61:459–466

Tsuchidi C, Richards FF, Ullu E (1986) The U2 RNA analogoue of *Trypanosoma brucei gambiense*: implications for a splicing mechanism in trypanosomes. Nucl Acids Res 14:8893–8903

Tyc K, Steitz JA (1989) U3, U8 and U13 comprise a new class of mammalian snRNPs localized in the nucleolus. EMBO J 8:3113–3119

Tyler BM (1985) Transcription of *Neurospora crassa* 5S rRNA genes requires a TATA box and three internal elements. J Mol Biol 196:801–811

Ulmanen I, Broni B, Krug RM (1983) Influenza virus temperature-sensitive cap (m^{7}pppNm)-dependent endonuclease. J Virology 45:27–35

Van Arsdell SW, Weiner AM (1984) Human genes for U2 small nuclear RNA are tandemly repeated. Mol Cell Biol 4:492–499

Van Santen VL, Spritz RA (1987) Nucleotide sequence of a bean (*Phaseolus vulgaris*) U1 small nuclear RNA gene: implications for plant pre-mRNA splicing. Proc Natl Acad Sci USA 84:9094–9098

Van Santen VL, Swain W, Spritz RA (1988) Nucleotide sequences of two soybean U1 snRNA genes. Nucl Acids Res 16:4176–4182

Vankan P, Filipowicz W (1988) Structure of U2 snRNA genes of *Arabidopsis thaliana* and their expression in electroporated plant protoplasts. EMBO J 7:791–799

Vankan P, Filipowicz W (1989) A U snRNA gene-specific upstream element and a –30 TATA box are required for transcription of the U2 snRNA gene of *Arabidopsis thaliana*. EMBO J 8:3875–3882

Vankan P, Edoh D, Filipowicz W (1988) Structure and expression of the U5 snRNA gene of *Arabidopsis thaliana*. Conserved sequence elements in plant U-RNA. Nucl Acids Res 16:10425

Waibel F, Filipowicz W (1990) Determination of RNA polymerase specificity of transcription of *Arabidopsis* U snRNA genes by promoter element spacing. Nature (Lond) 346:199–202

Waibel F, Vankan P, Filipowicz W (1990) The spacing between two cis promoter elements is a major factor determining the RNA polymerase specificity during transcription of U snRNA genes in *Arabidopsis thaliana*. J Cell Biochem 14B:171

Wasserman Da, Lee SI, Steitz JA (1989) Nucleotide sequence of HSUR 5 RNA from herpesvirus saimiri. Nucl Acids Res 17:1258

Watanabe-Nagasu N, Itoh Y, Tani T, Okano K, Koga N, Okada N, Oshima Y (1983) Structural analysis of gene loci for rat U1 small nuclear RNA. Nucl Acids Res 11:1791–1801

Weinberg RA, Penman S (1968) Small molecular weight monodisperse nuclear RNA. J Mol Biol 38:289–304

Weller P, Bark C, Janson L, Pettersson U (1988) Transcription analysis of a human U4C gene: involvement of transcription factors novel to snRNA gene expression. Genes Develop 2:1389–1399

Westin G, Zabielski J, Hammarstrom K, Monstein H-J, Bark C, Pettersson U (1984) Clustered genes for human U2 RNA. Proc Natl Acad Sci USA 81:3811–3815

Wise JA, Weiner AM (1980) *Dictyostelium* small nuclear RNA D2 is homologous to rat nucleolar RNA U3 and is encoded by a dispersed multigene family. Cell 22:109–118

Wise JA, Tollervey D, Maloney D, Swerdlow H, Dunn EJ, Guthrie C (1983) Yeast contains small nuclear RNAs encoded by single copy genes. Cell 35:743–751

Yu J-C, Nash MA, Santiago C, Marzluff WF (1986) Structure and expression of a second sea urchin U1 RNA gene repeat. Nucl Acids Res 14:9977–9988

Yuan Y, Reddy R (1988) Genes for human U3 small nucleolar RNA contain highly conserved flanking sequences. Biochim Biophys Acta 1008:14–22

Yuan Y, Reddy R (1989) Organization of spliceosomal U6 snRNA in the mouse genome. Mol Biol Rep 13:159–164

Yuan Y, Singh R, Reddy R (1989) Rat nucleolar 7-2 RNA is homologous to mouse mitochondrial RNase mitochondrial RNA-processing RNA. J Biol Chem 264:14835–14839

Yuo C-Y, Weiner AM (1989) A U1 small nuclear ribonucleoprotein particle with altered specificity induces alternative splicing of an adenovirus E1A mRNA precursor. Mol Cell Biol 9:3429–3437

Yuo C, Ares M Jr, Weiner AM (1985) Sequences required for 3' end formation of human U2 small nuclear RNA. Cell 42:193–202

Zagorski J, Tollervey D, Fournier MJ (1988) Characterization of an SNR gene locus in *Saccharomyces cerevisiae* that specifies both dispensible and essential small RNAs. Mol Cell Biol 8:3282–3290

Zeller R, Carri M-T, Mattaj IW, De Robertis EM (1984) *Xenopus laevis* U1 snRNA genes: characterisation of transcriptionally active genes reveals major and minor repeated gene families. EMBO J 3:1075–1081

Zhuang Y, Weiner AM (1986) A compensatory base change in U1 snRNA suppresses a 5' slice site mutation. Cell 46:827–835

Zieve GW, Penman S (1976) Small RNA species of the HeLa cell: metabolism and subcellular localization. Cell 8:19–31

Zieve G, Benecke BJ, Penman S (1977) Synthesis of two classes of snRNA species in vitro. Biochemistry 16:4520–4525

Zieve GW, Sauterer RA (1990) Cell biology of the snRNP particles. Crit Rev Biochem Mol Biol 25:1–46

Zieve GW, Sauterer RA, Feeney RJ (1988) Newly synthesized snRNAs appear transiently in the cytoplasm. J Mol Biol 199:259–267

Note added in proof:

Recently Mattaj and co-workers using the frog oocyte system (Vankan et al. 1990, EMBO J. 9:3397–3404) showed that the U6 snRNA transcripts do not leave the nucleus. In this respect, mepppG-capped U6 snRNA appears to be different from TMG-capped U1 and U2 snRNAs. The nuclear targeting signals present in U6 snRNA may be required for the re-accumulation of U6 snRNPs in the nucleus following mitosis in dividing cells. The pol III-transcribed plant U3 snRNA has an uncharacterized cap structure which is different from the TMG cap structure. Plant U3 snRNA lacking this cap structure can assemble into 15S monoparticles but not larger complexes involved in pre-rRNA processing (Kiss et al. 1991 Cell 65:517–526). These data suggest a role for the U3 snRNA cap structure in the assembly of large pre-rRNA complexes.

The DNA-Activated Protein Kinase, DNA-PK

T.H. CARTER[1] and C.W. ANDERSON[2]

1 Nuclear Protein Kinases

Cellular activity is controlled at numerous levels by protein phosphorylation (Hunter 1987). Energy metabolism is regulated by the concentration of ATP or its metabolic products through phosphorylation of the enzymes of glycogen metabolism (reviewed in Edelman et al. 1987). Dynamic cellular processes such as motility and division, integrative functions of the nervous and endocrine systems, and control of gene expression by intra- and extracellular signals depend on the reversible phosphorylation of key catalytic, structural, and regulatory proteins (for reviews, see Hunter and Cooper 1985; Hunter 1987; Edelman et al. 1987; Yardeen and Ullrich 1988; Murray and Kirschner 1989; Hunter 1989). For example, many hormone receptors (Yardeen and Ullrich 1988), oncogene and antioncogene products (Hunter and Cooper 1985; Bischoff et al. 1990), and transcription factors (Yamamoto et al. 1988; Prywes et al. 1988; Gould and Nurse 1989; Taylor and Young 1990; Lamph et al. 1990; reviewed by Hunt 1989) possess intrinsic protein kinase activity and/or are themselves regulated by phosphorylation. Activation of receptor-associated kinases may initiate a phosphorylation cascade within the cell by activating still other protein kinases that are their substrates (Morrison et al. 1989). Thus, protein kinases are important components of signal transduction pathways at the cell periphery, where physical or chemical signals interact with receptors to generate primary responses or second messengers, and also centrally, where nuclear architecture and biochemical activities are affected.

Nuclear protein kinases are undoubtedly key regulators of gene activity, growth, and cell division. Among the kinases purified from or localized to nuclei of eukaryotic cells, the two best-studied enzymes, variously called NI (Desjardins et al. 1972; Thornburg et al. 1978; Baydoun et al. 1981) and NII (Desjardins et al. 1972; Thornburg and Lindell 1977; Rose et al. 1981; Hara et al. 1981) or "nuclear casein kinases" (Zandomini and Weinmann 1984; Zandomini et al. 1986), have been isolated from a variety of species and tissues including HeLa cells. Neither enzyme is regulated by cyclic nucleotides, and both phosphorylate acidic domains of proteins in vitro, but the two enzymes differ in many other properties, including size and subunit composition (for a recent review, see Tuazon and Traugh 1990). Other

[1]Department of Biological Sciences, St. John's University, New York, NY 11439, USA
[2]Department of Biology, Brookhaven National Laboratories, Upton, NY 11973, USA

less well-characterized enzymes that may be localized to the nucleus include the product of the proto-oncogene *mos* (Watanabe et al. 1989), a chromatin-associated kinase which phosphorylates DNA-associated histones, especially histone H3 (Simpson 1981), a cGMP-dependent kinase which also phosphorylates histones preferentially (Hashimoto et al. 1979), a salt-sensitive enzyme(s) from HeLa cells (Quarless 1985; Friedrich and Ingram 1989), and possibly protein kinase C (Leach et al. 1989) and related enzymes (Ohno et al. 1988). The cAMP-dependent protein kinase has been found to accumulate in nuclei under certain conditions (Nigg et al. 1985; Sikorska et al. 1988). Ohtsuki et al. (1980a) identified a cAMP-dependent protein kinase in extracts of mouse spleen nuclei and partially purified a second, cyclic nucleotide-independent enzyme which is distinct from NII and phosphorylates histones and nonhistone chromosomal proteins better than acidic substrates such as casein. A cell cycle-regulated histone H1 kinase (Meijer and Pondaven 1988), otherwise known according to its genetic identity in yeast as the p34*cdc2* kinase, is the catalytic component of the cell cycle regulatory factor "MPF" (Arion et al. 1988; Labbe et al. 1988), which causes germinal vesicle breakdown (GVBD) in amphibian oocytes (Masui and Markert 1971) and is itself regulated by phosphorylation, as is the activity of at least one other component of the MPF complex (reviewed by Hunt 1989). The protein kinases that phosphorylate MPF in vivo have not yet been identified, although more than one enzyme is certainly involved, since components of MPF are phosphorylated on tyrosine, threonine, and serine residues (Draetta et al. 1988).

A multitude of nuclear functions and components are regulated by protein phosphorylation. Phosphorylation of nucleoplasmin increases during *Xenopus* oocyte maturation (Sealy et al. 1986), and the phosphorylation state of lamins A and C is correlated with their changes in organization during mitosis (Gerace 1986), perhaps induced by a lamina-associated protein kinase (Dessev et al. 1988). Protein kinase inhibitors prevent induction of GVBD in amphibian oocytes (Dufresne et al. 1989), whereas activation of protein kinase C induces GVBD in *Spisula* oocytes (Eckberg et al. 1987). The *cdc2* kinase phosphorylates several substrates in vitro at the same sites as are naturally phosphorylated in vivo: these include lamin B, histone H1, nucleolin, and pp60[src] (reviewed by Moreno and Nurse 1990) as well as SV40 T-antigen (McVey et al. 1989) and the cellular anti-oncogene product p53 (Bischoff et al. 1990). The *mos* protein, a serine/threonine protein kinase, has recently been shown to be the "cytostatic factor" that interacts in a regulatory fashion with MPF (Watanabe et al. 1989). A number of other growth-regulating factors such as the cellular retinoblastoma susceptibility product are also phosphorylated at specific times during the cell cycle (DeCaprio et al. 1989).

2 Nucleic Acid Effects on Protein Kinase Activity

DNA is usually thought of as the repository of genetic information and as a target for site-specific recognition by enzymes that utilize DNA as a substrate (e.g., nucleases and helicases) or as a template (e.g., DNA and RNA polymerases) and by various proteins that regulate the action of these enzymes. Rarely has DNA itself been thought of as a regulatory element for enzyme activity. However, the in vivo substrates for nuclear protein kinases are likely to include structural and regulatory proteins associated with chromatin, and it is therefore reasonable that DNA might modify the activity or substrate specificity of these kinases. For example, many proteins that participate in transcription and DNA replication are known to interact with specific DNA sequences, and kinases that regulate the activity of these proteins might interact with the same DNA sequences or with other regions of the DNA juxtaposed with the *cis*-acting regulatory sequences in vivo. This would constitute an especially efficient arrangement in the case of DNA-bound proteins that are subsequently inactivated by phosphorylation either to cause dissociation or to prevent reformation of the complex after equilibrium dissociation of the unphosphorylated species. The existence of a DNA-activated protein kinase thus provides a potential feedback mechanism whereby the effects of cellular functions on DNA structure and activity could modulate those same functions.

Apart from DNA-PK, we know of few protein kinases whose activities are directly regulated by nucleic acids. DNA was reported to inhibit casein phosphorylation by the NII kinase (Rose et al. 1981), but it is currently unclear whether this inhibition serves a physiological role. A double-stranded RNA-activated protein kinase, DAI (Sen et al. 1978), specifically phosphorylates the translation factor EIF-2a (Farrell et al. 1977, 1978). A cyclic nucleotide-independent enzyme isolated from mouse spleen nuclei (Ohtsuki et al. 1980a,b, 1982) was stimulated up to five-fold by DNA specifically for phosphorylation of two small nonhistone proteins, and histone phosphorylation by the chromatin-associated cGMP-dependent histone kinase could also be stimulated by DNA (Hashimoto et al. 1979); in both cases the effect was thought to result from interaction of the substrates with DNA. However, only one protein kinase has been shown to require DNA as a general cofactor: a kinase capable of DNA-activated protein phosphorylation in HeLa cells, reticulocyte lysates, and oocytes from *Xenopus* and sea urchin which was independently discovered by our groups (Walker et al. 1985; Carter et al. 1988).

3 Detection of DNA-Stimulated Protein Phosphorylation in Cell Extracts

Anderson and colleagues discovered that several nucleic acid preparations induced the phosphorylation of 90 kDa and 45 kDa polypeptides in rabbit reticulocyte lysates (Walker et al. 1985). These polypeptides were not phosphorylated by the double-stranded RNA-activated protein kinase, DAI. Pure double-stranded RNA (dsRNA) at the low concentrations which activate DAI did not induce phosphorylation of the 90 kDa and 45 kDa polypeptides, and high concentrations of dsRNA which inhibit DAI activation and eIF2a phosphorylation did not inhibit phosphorylation of these polypeptides (Fig. 1). Phosphorylation of the 90 kDa and 45 kDa polypeptides was first noted with tobacco mosaic virus (TMV) RNA and with commercial preparations of tRNA, but a number of other RNA preparations, including

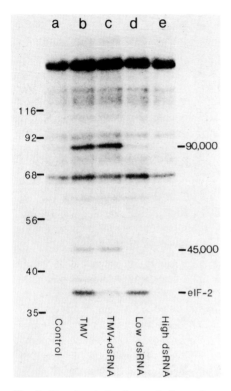

Fig. 1. Phosphorylation in a reticulocyte lysate in response to TMV RNA and double-stranded RNA. Extracts were pre-incubated at 30 °C for 10 min prior to addition of γ-^{32}P-ATP. Labeling was stopped after a further 10 min by addition of 5 vol. SDS sample buffer. After SDS-PAGE the stained and dried gel was autoradiographed. The autoradiograph shows lysate containing *a* no additons; *b* TMV RNA added at 100 µg/ml; *c* TMVRNA at 100 µ/ml + ds RNA at 100 µg/ml; *d* dsRNA alone at 100 µg/ml; *e* dsRNA alone at 100 µg/ml. The positions of molecular weight markers are indicated at the *left* of the figure; the positions of the 90 000 M_r, the 45 000 M_r and the α-subunit of eIF-2 (38 000 M_r) are indicated at the *right*

poliovirus RNA and globin mRNA, were not phosphorylation effectors. That phosphorylation was induced not by RNA, but by dsDNA, was suggested by sensitivity to boiling but not to heating at 90 °C, and by sensitivity to DNase I and DNase II but not to RNase A. Subsequently, a variety of natural and synthetic dsDNAs, including calf thymus, bacteriophage lambda, pBR322, bacteriophage M13 replicative form, poly (dG•dC) and poly (dA•dT) were found to induce phosphorylation, whereas pure RNAs and ssDNA, including boiled calf thymus DNA, either strand of DNA from M13, poly (dT) and poly (dC), were not effective. Similar DNA-dependent phosphorylation was then found in extracts from a variety of species, including frog, clam and sea urchin oocytes (Fig. 2), and cultured human cells (Walker et al. 1985). The 90 kDa polypeptide phosphorylated in human, rabbit and *Xenopus* extracts was identified as the 90 000 Da heat-shock protein, *hsp*90. However, because the phosphorylation state of *hsp*90 in vivo was not known at the time,

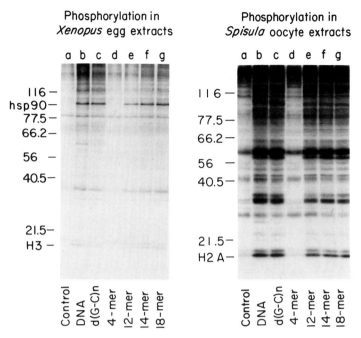

Fig. 2. Activation of *Xenopus* and *Spisula* dsDNA-dependent kinase by synthetic ds oligonucleotides. Extracts were obtained from R.A. Laskey (*Xenopus*) and T. Hunt (*Spisula*), Cambridge University. Experimental conditions were similar to those described in Fig. 1. Poly(G·C)n indicates an annealed mixture of commercially prepared homopolymers with average chain lengths greater than 30 nucleotides. The positions of several molecular weight markers and of *hsp*90 are indicated. Although several polypeptides in *Spisula* exhibit dsDNA-dependent phosphorylation, none of these correspond in mobility to human, rabbit or *Xenopus* hsp90. (Anderson and Walker unpubl.)

Walker et al. could not distinguish whether DNA activated a kinase activity or a phosphatase activity that allowed labeling of *hsp*90 through turnover.

While studying the phosphorylation of transcription complexes in an in vitro system derived from HeLa cells, Carter et al. (1988) independently discovered that dsDNA induced the phosphorylation of several endogenous substrates. The molecular weights of these substrates were similar to those observed in the reticulocyte system. The effect of DNA on phosphorylation was seen in both whole cell and nuclear extracts, and was specific for linear dsDNA, since neither supercoiled nor denatured DNA caused phosphorylation of the characteristic set of polypeptides when added to extracts. When either nuclear or whole cell extracts were fractionated by ion exchange chromatography, a preparation was obtained in which phosphorylation of exogenously added α-casein was stimulated five- to tenfold by DNA. DNA from a variety of sources was effective at activating phosphorylation, as long as the polynucleotide was double-stranded. Single-stranded DNA was found to be inhibitory, and activity could be removed from crude extracts by addition of DNA and high speed centrifugation, suggesting that the putative kinase could physically bind to DNA.

4 Purification of DNA-PK from HeLa Cells

The purification of DNA-PK utilizes ion exchange, gel filtration, and affinity chromatography steps, some of which were adapted from procedures for purification of other nuclear protein kinases (Ohtsuki et al. 1980a; Rose et al. 1981; Zandomini and Weinmann 1984; Zamdomini et al. 1986). Stimulation of phosphorylation of either α-casein or purified *hsp*90 by dsDNA is used as the specific assay (Carter et al. 1990; Lees-Miller et al. 1990). The procedures utilized in our laboratories have a number of steps in common and result in purified preparations with many similar properties; the differences observed between our enzymes may be ascribed to differences in extract preparation and specific chromatographic procedures. In one case, cells are disrupted by shearing in hypotonic buffer, and crude nuclei obtained by low speed centrifugation are extracted by high salt prior to ion exchange chromatography (Dignam et al. 1983; Carter et al. 1988, 1990). In the other, whole cells are disrupted by freezing and thawing, and differential centrifugation is used to isolate a subcellular particulate fraction, which is then extracted with high salt prior to chromatographic fractionation (Lees-Miller et al. 1990). The DNA-PK solubilized by high salt in each case may associate with different proteins present in the two kinds of extracts, and these associations may influence the subsequent purification efficiency and biochemical behaviour of DNA-PK. Table 1 summarizes a typical purification from nuclei.

The two procedures each yield preparations that are probably >1000-fold purified by either recovery of total protein or specific activity. The figures for both recovery and extent of purification are subject to the uncertainties inherent in

Table 1. Purification of the DNA-dependent protein kinase

Purification step	Total activity[a] (units × 10^{-3})	Stimulation by DNA[b]	Total protein (mg)	Specific activity (units/mg) × 10^{-3}	Purification factor[c]
Cytoplasmic S-100	5.6	1.1	243	0.02	—
Nuclear extract	119.5	1.4	374	0.32	1.6
(NH₄)₂SO₄ precipitate	58.2	1.7	219	0.27	1.4
DEAE-Sephacel	36.7	2.6	3.36	10.9	54
Phosphocellulose	61.2	62.9	1.14	53.7	268
Phosvitin-Sepharose	60.7	24.2	0.19	316.0	1580
DNA-Cellulose	24.5	23.5	0.09	286.8	1434
Sephacryl S-300HR	18.9	>100[d]	0.02	943.5	4718

[a]Activity in the presence of 50 µg/ml sonicated salmon DNA minus activity without added DNA. 1 unit = 1 pmol ^{32}P transferred from [γ-^{32}P]ATP to casein at 30 °C in 1 min.
[b]Activity in the presence of DNA divided by activity without DNA.
[c]The total DNA-dependent activity in cell lysates was calculated by adding the total activities in the cytoplasmic S-100 and nuclear extracts and dividing by total protein. This value, 0.20 units/mg, was taken as a purification factor of 1.0.
[d]DNA-independent activity was not significantly different from 0-time control.

measuring activity in crude extracts and in quantitating small amounts of pure protein, and are also complicated by the fact that the enzyme appears to lose activity easily during purification. Nevertheless, we estimate that DNA-PK constitutes as much as 0.1% of the total nuclear protein extractable in 0.4 M KCl. The specific activity of purified preparations is typically between 200 and 900 nmol of phosphate transferred to an exogenous substrate (denatured α-casein, at 1–2 mg/ml, or purified *hsp*90 at 0.5 mg/ml) per min per mg protein at either 37 °C (Carter et al. 1990) or 30 °C (Lees-Miller et al. 1990). This activity is within the range reported for highly purified preparations of other nuclear protein kinases, for example casein kinase NII. From 50 g of HeLa cells it is possible to obtain several hundred micrograms of nearly homogenous enzyme using either procedure (Carter et al. 1990; Lees-Miller et al. 1990).

The stimulatory effect of DNA on DNA-PK activity typically increases during purification (Table 1). Although the most highly purified preparations occasionally have no detectable casein phosphorylating activity in the absence of added DNA, most phosphorylate casein at up to 5% of the rate that is observed in the presence of DNA. Digestion with DNaseI reduces this activity, but sometimes does not eliminate it. Because phosphorylation of *hsp*90 in the absence or presence of added DNA was shown to occur at identical sites (Lees-Miller and Anderson 1989b), it is probable that the enzyme possesses a basal level of DNA-independent activity. However, this activity is minor compared to the level seen in the presence of DNA.

5 Physical Characteristics of DNA-PK

Purified preparations obtained from nuclei contain a high M_r polypeptide as the only major component on reducing SDS-PAGE, with other large polypeptides sometimes detectable by silver staining as minor components (Carter et al. 1990). Purified enzyme from the cytoplasmic particulate fraction also contains a high M_r major component that co-migrates on gels with the major nuclear-derived component. At early stages of purification, the cytoplasmic preparation contains several other prominent DNA-binding polypeptides, at least four of which are substrates for DNA-PK (Lees-Miller et al. 1990). Although we refer to the major high M_r band as the 300 kDa polypeptide, its apparent molecular weight is estimated to be between 300 and 350 kDa; a precise value for a protein this large is difficult to obtain because the M_r calculated from electrophoretic mobility depends on the gel system, electrophoresis conditions, and protein standards employed. Use of nuclei as the starting material and shallow gradient elution with KCl during the initial chromatographic steps result in fewer minor bands in the final preparation, and show a kinase activity profile of fractions from the final Mono Q step that closely corresponds to the staining intensity of the 300 kDa polypeptide (Carter et al. 1990). In preparations starting from cells disrupted by freezing and thawing, the peak of kinase activity sometimes elutes slightly ahead of the 300 kDa peak from columns containing immobilized single-stranded DNA and slightly after the 300 kDa peak during chromatography on MonoS (Lees-Miller et al. 1990), but this might be due to the presence of inactive forms of the enzyme which still retain the ability to bind DNA, albeit less strongly than active enzyme.

When DNA-PK preparations isolated by the two methods were autophosphorylated in the presence of γ-^{32}P-ATP and DNA, the labeled 300 kDa polypeptides from each preparation co-migrated on SDS-PAGE, and partial digestion of these two polypeptides with *Staphylococcal* V8 protease gave identical phosphopeptide patterns (Fig. 3), indicating that the two high M_r species are identical at the level of resolution of the method. We have further determined that the enzymes isolated by our two laboratories have similar substrate specificities in vitro, and that *hsp*90 is phosphorylated at the same N-terminal threonine residues by both preparations (Carter et al. 1989; see below).

Independent identification of a 300 kDa polypeptide as the major or sole component of DNA-PK is provided by the observation that one out of three monoclonal antibodies which recognize this polypeptide on Western blots (MAb18–2) also inhibits DNA-activated casein phosphorylation by more than 50%, and quantitatively immunoprecipitates DNA-PK activity (Carter et al. 1990). All three antibodies also recognize extensive sets of smaller polypeptides on Western blots when enzyme preparations at various stages of purification are overloaded on the gels and conditions of detection are chosen for maximum sensitivity. Because individual polypeptides comprising the set recognized by each MAb are present in the same proportion at early stages of purification as in highly purified enzyme, it appears likely that the 300 kDa polypeptide suffers limited endopeptidase cleavage either in vivo or during extract preparation, and that the cleavage products co-purify with intact DNA-PK until they are separated under reducing and denaturing conditions. The existence of these cross-reacting polypeptides probably accounts for at least some of the minor silver-staining bands present in purified preparations. Assuming that the 300 kDa

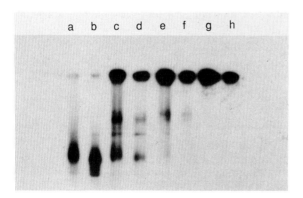

Fig. 3. Partial V8 protease digestion of autophosphorylated 300 kDa polypeptides. Partially purified DNA-PK preparations from cytoplasm (lanes *a,c,e,* and *g*) and nuclei (lanes *b,d,f,* and *h*) were autophosphorylated with γ-^{32}P-ATP, run on SDS-PAGE and autoradiographed; the 300 kDa band was excised and digested with 50, 5, 1 or µg/ml Staphylococcal V8 protease (lanes *a* and *b, c* and *d, e* and *f, g* and *h*, respectively) for 20 min before electrophoresis through 12.5% SDS-PAGE and autoradiography

polypeptide is the intact, enzymatically active component, the relative elution volume of DNA-PK on gel exclusion columns indicates that the enzyme exists in extracts at or above 0.3 M KCl as a monomer of approximate MW = 300 000 (Carter et al. 1990; Lees-Miller et al. 1990).

6 Subcellular Localization

Although DNA-PK has been purified from the cytoplasmic particulate fraction after hypotonic disruption of HeLa cells and was also identified in reticulocyte lysates (Walker et al. 1985), the enzyme is not predominantly cytoplasmic in vivo. On one hand, it is known that nuclear proteins can diffuse into the cytosol after cell disruption (Paine et al. 1983) where they may bind to organelles under hypotonic conditions. However, the crude preparations of nuclei from which DNA-PK has also been purified contain cytoplasmic and mitochondrial contaminants, based on microscopic examination and assays for marker enzymes (Carter et al. 1990). We therefore approached the question of subcellular localization by (1) biochemical analysis of more highly purified nuclei and (2) by immunocytochemical staining using several different monoclonal antibodies. Both procedures showed that the preponderance of DNA-PK is localized in the nucleus of HeLa cells growing in monolayer (Carter et al. 1990). Nuclear staining was also observed in monolayer cells from human breast and lung carcinomas, and in a human diploid fibroblast line (Carter et al., unpubl.). In HeLa cells, intense immunofluorescent staining of what appeared to be condensed chromosomes was occasionally observed, suggesting that DNA-PK may associate with chromatin during mitosis. Variability in staining intensity of nuclei suggests that the amount or accessibility of the reactive epitope may differ among various cells in a single population, a result commonly observed with other nuclear accumulating proteins.

DNA-PK is also present in amphibian nuclei. Activity was detected in *Xenopus* germinal vesicles isolated by standard aqueous methods (Anderson et al., unpubl.). Furthermore, a 300 kDa polypeptide which cross-reacts with an anti DNA-PK monoclonal antibody on Western blots is present in oocyte nuclei which are extruded under oil (Carter and Paine, unpubl.), a method which isolates nuclei with minimal cytoplasmic contamination and assures that they retain their in vivo protein constituents (Lund and Paine 1990).

7 Phosphate Donor and Cofactor Requirements

Cyclic 3'–5' AmP or cyclic 3'–5' GMP have no effect on DNA-activated phosphorylation of casein. The abilities of γ-^{32}P-labeled ATP and GTP to act as phosphate donors were compared by Lineweaver-Burk analysis, using casein as substrate in the presence of DNA, and a partially purified DNA-PK preparation (through the phosvitin-Sepharose step illustrated in Table 1). ATP was preferred (K_m^{app} = 41 μM and V_{max} = 108.9 nmoles min^{-1} mg^{-1}), but GTP was also used (K_m^{app} = 176 μM and

$V_{max} = 17.4$ nmoles min^{-1} mg^{-1}). Kinase obtained from cytoplasmic extracts did not use GTP to a significant extent as a phosphate donor, as determined by inability of GTP to compete with ATP in the phosphorylation of hsp90 (Lees-Miller et al. 1990). It is not clear whether weak utilization of GTP by the nuclear enzyme represents a significant difference between the two preparations, or whether the apparent discrepancy is the result of differences in assay conditions or in the ways in which GTP utilization was measured.

Among the divalent metal ions tested, Mg^{2+} (optimum concentration 6–10 mM) and Mn^{2+} (2.5 mM) were preferred co-factors for DNA-activated kinase activity with casein as the receptor and ATP as the phosphate donor, but the rate of phosphorylation in the presence of Mg^{2+} was higher than with Mn^{2+}. Ca^{2+} was a weak activator, CO^2 and Zn^{2+} were inactive, and EDTA completely inhibited Mg^{2+}-activated phosphorylation. Using 7.5 mM MgCl$_2$ as the standard condition, DNA-activated phosphorylation was greatest at low monovalent cation concentrations (around 50 mM), and inhibited by KCl at concentrations >150 mM (Carter et al. 1990; Lees-Miller et al. 1990). Sensitivity to salt concentrations in the physiological range may be an artifact of isolation, however, considering that interaction of the enzyme with nuclear DNA in vivo would almost certainly involve other chromatin proteins in a highly compartmentalized biochemical environment. For example, a similar effect of monovalent cation concentration is also seen on in vitro transcription (Lewis and Burgess 1982). Nevertheless, the inhibition of DNA-PK by salt in vitro further distinguishes this enzyme from most other protein kinases.

8 Effects of Inhibitors

DNA-PK has a unique pattern of sensitivity to many inhibitors of protein kinases (Table 2). Spermine, which stimulates casein kinase II and inhibits casein kinase I

Table 2. Effect of protein kinase inhibitors on DNA-PK activity

Inhibitor	Concentration for 50% inhibition[a]	
N-ethylmaleimide	0.7	mM
Heparin	0.8–1.0	µg/ml
Spermine	1	mM
Spermidine	10	mM
2-aminopurine	4	mM
N,N-dimethylaminopurine	0.5	mM
N,N-dimethylaminopurine riboside	5	mM
Inorganic phosphate	20	mM
Pyrophosphate	3	mM
ADP, AMP	>300	µM
DRB, quercetin, rutin	>100	µM

[a]The symbol ">" indicates 50% inhibition was not obtained at the concentration given.

(Verma and Yu Chen 1986), is a potent inhibitor of PKD's casein phosphorylation activity. Heparin, which inhibits casein kinase II completely at 0.1 µg/ml requires a tenfold higher concentration to inhibit the DNA-activated kinase (Rose et al. 1981; Hara et al. 1981; Zandomini et al. 1986). N-ethylmaleimide, which does not inhibit most nuclear protein kinases, does inhibit the DNA-stimulated enzyme in a dose-dependent manner, indicating that SH groups are essential for activity. 2-Aminopurine and $N^{6,6}$ dimethylaminopurine, which inhibit a number of protein kinases (Ohtsuki et al. 1980a; Rose et al. 1981), also inhibit DNA-dependent casein phosphorylation, but 5,6-dichloro-1-β-D-ribofuranosylbenzimidazole (DRB), quercetin, and rutin do not inhibit DNA-PK extensively at concentrations that completely abolish activity of casein kinase II (Zandomini and Weinmann 1984; Zandomini et al. 1986). Low concentrations of nonionic detergents (0.1% or less) do not inhibit activity significantly (Lees-Miller et al. 1990), and inclusion of 0.02% Tween-20 in buffers during purification has been found to improve both yield and purity (Carter et al. 1990). The enzyme is strongly inhibited by phosphate and pyrophosphate, but not by ADP or AMP (Lees-Miller et al. 1990). Single-stranded DNA is also inhibitory.

9 Substrate Specificity

9.1 Autophosphorylation

In pure preparations containing only the 300 kDa polypeptide as the major component, incubation with Mg2+, γ-^{32}P-ATP and DNA results in a single labeled band after electrophoresis and autoradiography that co-migrates with silver-stained DNA-PK (Carter et al. 1990); this species is also radiolabeled in less pure preparations, along with several of the minor co-fractionating polypeptides (Less-Miller et al. 1990). Alkaline phosphatase removes the label from the 300 kDa polypeptide, and addition of competing protein substrates to the reaction, but not BSA, also reduces labeling. Thus DNA-PK is capable of autophosphorylation in the presence of DNA, although it is not known whether this represents an intra- or and intermolecular event. The fact that autophosphorylation also inhibits the kinase activity of DNA-PK with respect to both *hsp*90 (Lees-Miller et al. 1990) and casein (Carter et al. 1990) substrates suggests that autophosphorylation may be important for regulating the enzyme's activity in vivo, particularly if it is intramolecular, or if DNA-PK binds to DNA in multiple copies, e.g., as a dimer. If DNA-PK activity is regulated in this way, activation of the kinase by DNA would be a short-lived event in the absence of competing substrates. Autophosphorylation could inhibit kinase activity by causing either dissociation of the enzyme-DNA complex, or conformational changes that prevent substrate binding or catalysis. We do not yet know whether DNA-PK is phosphorylated in vivo.

9.2 Substrate Preference and Phosphorylation Sites

We have tested a number of different proteins for their ability to act as acceptors for DNA-PK phosphorylation, usually by incubation with the enzyme under standard reaction conditions followed by gel electrophoresis and either autoradiography or scintillation counting of the excised polypeptide band. Among the usual protein kinase substrates tested at high concentrations (~50 μM), purified lysine-rich histone fraction H1 is phosphorylated weakly by highly purified enzyme in the presence of DNA (Carter et al. 1990). Other histones purified from calf thymus were not phosphorylated in the presence or absence of DNA by either enzyme preparation. However, in frog oocytes, one of the polypeptides that is phosphorylated in a DNA-dependent manner co-migrates on SDS/PAGE with a maternal form of histone H2A (Anderson, Walker, and Laskey unpubl.) which contains a potential phosphorylation site for DNA-PK (see below). At low protein concentrations (e.g., 250 μg/ml), denatured α-casein was not significantly phosphorylated without added DNA, but was the best substrate for highly purified enzyme per unit mass in the hands of one of us (Carter et al. 1990), followed by SV40 T-antigen (cloned in baculovirus) and α-*hsp*90. Denatured phosvitin and BSA were phosphorylated weakly. Only threonine and serine were labeled by DNA-activated phosphorylation of α-casein in the presence of γ-[^{32}P]ATP, and serine was the only amino acid phosphorylated in SV40 T-antigen and p53 (Lees-Miller et al. 1990; see below).

Substrate preference was different for enzyme prepared from whole cells, in that SV40 T-antigen was the best substrate tested, followed by α-*hsp*90. Casein was not phosphorylated as well as the former two proteins, and phosphorylation of BSA and histones was not observed (Lees-Miller et al. 1990). Explanations for the difference in substrate preference include the presence of additional proteins in the whole cell enzyme preparation, use of native β-casein as opposed to denatured α-casein, variability in endogenous phosphorylation of baculovirus-produced T-antigen and, perhaps most importantly, the fact that enzyme prepared in our two laboratories was purified in one case according to its optimum ability to phosphorylate α-casein and, in the other, to its ability to phosphorylate α-*hsp*90. Thus, different forms of the kinase, either initially associated with different endogenous proteins, or perhaps containing different degrees of posttranslational modification (such as phosphorylation), may have been selectively purified by the two methods.

*Hsp*90 was the first endogenous substrate for DNA-PK identified in extracts from whole cells (Walker et al. 1985). The human protein is present in two forms, α and β, which differ in their N-terminal sequence (Hickey et al. 1989). Only the α form was phosphorylated by DNA-PK, and sequence analysis showed that two unique threonine residues in the N-terminal sequence NH₂-PEETQTQDQPME were phosphorylated (Lees-Miller and Anderson 1989b). This site was not phosphorylated by casein kinase II (Lees-Miller and Anderson 1989a). Although very little phosphothreonine was found on α-*hsp*90 isolated from growing HeLa cells, the phosphorylation of the unique site by DNA-PK in vitro afforded an opportunity to search the protein sequence databases for other potential substrates. Among the eukaryotic proteins containing the sequence motif E(T/S)Q and Q(T/S)Q +

D(T/S)Q, several are known to bind DNA (or to be physically associated in vivo with other proteins that do), including SV40 T-antigen, the human progesterone receptor, the p53 anti-oncogene product, the generalized transcription factor Sp1, the enhancer-binding proteins Oct-1 and Oct-2, and the nuclear auto-antigen Ku. With the exception of the progesterone receptor, all of these proteins can be phosphorylated by DNA-PK in vitro in a DNA-dependent manner; the phosphorylated sites and the biological significance of such phosphorylation are not yet determined. Among the in vitro substrates that co-purify with DNA-PK, sequence analysis of the purified 81 kDa and 61 kDa polypeptides (which purify as a complex) identifies these as components of the Ku autoantigen (Mimori et al. 1986; Lees-Miller et al. 1990).

   In terms of potential in vivo substrates and their significance, a recent provocative result involves the case of the transcription factor Sp1. Phosphorylated and unphosphorylated forms of Sp1 can be distinguished by their electrophoretic mobilities, and Sp1 that is engaged in transcription activation appears to consist predominantly of the phosphorylated species (Jackson et al. 1990). Phosphorylation of Sp1 by purified DNA-PK induces an electrophoretic mobility shift similar to that observed during Sp1-promoted transcription in vitro (Jackson et al. 1990), indicating that DNA-PK may be involved in regulation of Sp1-dependent transcription. This conclusion is strengthened by the fact that Sp1 is not a good substrate unless it is bound to its cognate G/C-rich enhancer sequence; other DNA sequences, which activate DNA-PK for phosphorylation of hsp90, for example, do not activate phosphorylation of Sp1. It is not known whether this additional level of specificity results from a conformational change in Sp1 induced by binding to DNA, an increased ability of DNA containing an Sp1-binding sequence to activate DNA-PK, or is simply a matter of bringing the substrate, which is present at very low concentration, into physical proximity with kinase, which binds to the same DNA molecule as a result of its generalized affinity for DNA. It is possible that all three explanations could account for the highly efficient Sp1 phosphorylation observed by Jackson et al. However, the facts that poly(dG•dC) was the best synthetic activator of DNA-PK (Lees-Miller et al. 1990) and that an oligonucleotide containing an Sp1-binding sequence had the lowest K_a among several enhancer sequences tested for activation of casein phosphorylation (Carter et al. 1990) argue that DNA sequence may be important for activation of the enzyme, in addition to its direct effect on a substrate such as Sp1 through formation of a DNA-activated substrate complex.

10 Effects of Polynucleotides

The requirement for double-stranded DNA (dsDNA) is the unique characteristic of DNA-PK as a protein kinase. Purified enzyme preparations are highly dependent upon addition of native salmon DNA for phosphorylation of α-casein and hsp90, as well as other exogenous protein substrates (20-fold to >100-fold stimulation). Synthetic, double-stranded polydeoxyribonucleotides such as poly (dI•dC), poly (dG•dT) are efficient activators at high concentration (>10 µg/ml), whereas highly purified preparations of tRNA and single and double-stranded viral RNAs, as well as synthetic polynucleotides such at poly (dI•rC) and poly (rI•rC) are inactive (Carter et al. 1990; Lees-Miller et al. 1990). Single-stranded polynucleotides such as poly (dT) and poly (dC) are also inactive (Lees-Miller et al. 1990), and heat-denatured calf thymus DNA is inhibitory (Carter et al. 1988; Carter et al. 1990).

Activation of DNA-PK by DNA is specific for linear, double-stranded polydeoxyribonucleotides. The relative inability of supercoiled DNA preparations to activate DNA-PK was observed in both purified enzyme preparations and crude cell extracts (Carter et al. 1988; Carter et al. 1990; Lees-Miller et al. 1990). To assess the possible importance of the structure of free ends of dsDNA for DNA-PK activation, different restriction endonucleases, as well as sonication, were used to generate DNA fragments, and the resultant DNA preparations were also treated with a single-strand-specific nuclease (S1) to generate flush-ended DNA molecules (Carter et al. 1990). None of these treatments substantially changed the ability of each DNA preparation to activate the enzyme, indicating that neither short terminal regions of single-stranded DNA nor specific terminal nucleotide sequences are required for activation. The finding that supercoiled plasmid DNA was a poor activator compared to the same DNA that had been linearized with restriction endonucleases, or intentionally nicked by freezing and thawing, suggests that the enzyme needs tortionally unconstrained DNA to effect productive binding, and is consistent with binding in vivo to non-nucleosomal regions of DNA such as those undergoing transcription or replication.

Although there is no direct evidence that DNA activates DNA-PK by binding to the enzyme, rather than to the substrate, as proposed for the two other instances of protein kinase activation by DNA (Hashimoto et al. 1979; Ohtsuki et al. 1982), several experiments show that the enzyme interacts directly with DNA. Attempts to detect co-sedimentation of purified enzyme and activating DNA on glycerol gradients were frustrated by an inability to recover enzyme activity from the gradients. We subsequently found this to be the result of enzyme inactivation by high glycerol concentrations. Ultraviolet cross-linking experiments show that the 300 kDa polypeptide can be covalently linked to short end-labeled oligodeoxyribonucleotides (Lees-Miller et al. 1990), and thermal inactivation of DNA-PK is more rapid in the presence of dsDNA (Carter et al. 1990). Purified DNA-PK also retards the electrophoretic mobility of synthetic oligonucleotides in a manner that suggests the formation of structurally discrete complexes (Carter et al. 1990). Competition gel-shift experiments indicate that binding of purified DNA-PK to DNA is relatively independent of nucleotide sequence under conditions that maximally activate kinase activity in vitro (Carter et al. 1990). This result is consistent with the observation that activation of DNA-PK is also independent of DNA sequence (Lees-Miller et al. 1990; Carter et al. 1990). However, even under the low salt conditions used in the assay of enzyme activity, which would be expected to increase nonspecific protein-DNA interaction, the apparent K_a's for different DNAs varies by several orders of magnitude (Carter et al. 1990), whereas the calculated V_{max} values, which are close to the rates determined by routine enzyme assay at DNA excess (50 μg/ml), are within a factor of 4. Results of these kinetic studies thus leave open the question of potential sequence specificity. Conditions more closely approximating those which might be expected to occur in vivo must be examined, including effects of other proteins that might bind to DNA in a sequence-specific manner, as in the case of Sp1.

11 Occurrence of DNA-PK in Other Cells

Indirect immunofluorescence using monoclonal antibodies against DNA-PK gave intense nuclear staining of a number of human cells growing in culture, including the adenovirus-transformed fetal kidney cell line 293, tumor cell lines from lung, breast, and pancreatic carcinomas, and diploid human fibroblasts. Thus, it is likely that the enzyme is generally expressed in human cells. Nuclear immunofluorescence was also observed in monolayer cultures of simian cells (Cos cells) and in chemically transformed mouse pancreatic acinar cells, although enzymatic activity has not been assayed in cell-free extracts from these species. Walker et al. (1985) were unable to detect DNA-stimulated protein phosphorylation in extracts from murine L-929 cells, chick embryo fibroblasts, or baby hamster kidney cells, but it is possible that extraction or assay conditions were not optimal for enzyme from these sources. In contrast, the presence of a DNA-PK-like activity was reported in whole cell extracts from *Xenopus, Arbacia,* and *Spsisula* oocytes (Walker et al. 1985; Fig. 2), and a high M_r polypeptide was detected on Western blots of *Xenopus* nuclei using a monoclonal antibody specific for the 300 kDa HeLa polypeptide (Carter and Paine, unpubl.). Thus, unless the oocytes of nonmammalian species are a special case, the occurrence of DNA-PK in proliferating human cells and in such distantly related species as *Xenopus* and sea urchin suggests that the enzyme is widely distributed in animals.

12 Comparison with Other Nuclear Protein Kinases

DNA-PK is similar to a number of other nuclear protein kinases in not requiring cyclic nucleotides, in its specificity for phosphorylation of threonine and serine residues, and in its ability to undergo autophosphorylation. However, DNA-PK has unique physical and biochemical characteristics that distinguish it from other known nuclear protein kinases. (1) The combination of large size of the native enzyme (approximately 300 kDa) and monomeric structure are unique among serine-threonine protein kinases: NI was reported to exist as a monomer (Baydoun et al. 1981) or homodimer (Thornburg et al. 1978) of 25 kDa subunits, and the mouse spleen enzyme (Ohtsuki et al. 1980a) as a monomer of 45 kDa, whereas NII (Tuazon and Traugh 1990) and a salt-sensitive HeLa kinase (Quarless 1985) are multi-subunit enzymes composed of polypeptides in the range of 20 kDa to 60 kDa. (2) Although several receptors for polypeptide growth hormones possess protein kinase activity and have molecular masses approaching that of DNA-PK (Yardeen and Ullrich 1988), all are tyrosine-specific, and none has been found in association with cell nuclei. (3) Where comparative data are available, the affinity of DNA-PK for ion exchange resins is different from other nuclear protein kinases. (4) Other data reported here which further distinguish DNA-PK from known nuclear protein kinases are its inhibition by KCl concentrations above 0.15 M, inhibition by NEM and spermine, and activation by DNA. The dependence of DNA-PK on dsDNA for phosphorylation of casein distinguishes this enzyme from the mouse spleen kinase, which phosphorylates casein and other exogenous substrates efficiently in the absence of added DNA (Ohtsuki et al. 1980a,b). It seems unlikely that Ohtsuki's result could have been due to contamination of his preparations with DNA, since phosphorylation of the specific nonhistone substrates could be stimulated as much as fivefold by added DNA, whereas DNA had no effect on phosphorylation of other substrates tested, including casein and phosvitin. A salt-sensitive nuclear casein kinase that has been partially purified from HeLa nuclei by Friedrich and Ingram (1989) has properties similar to DNA-PK, including a high M_r polypeptide. This enzyme may be related or identical to DNA-PK, but no effect of DNA was reported.

13 Conclusions and Future Directions

We have documented the existence of a novel protein kinase, which we call DNA-PK. The enzyme is predominantly nuclear in vivo, and has several unique properties, including large molecular size and a generalized stimulation by dsDNA. This last property implies potential roles for DNA-PK in the regulation of nuclear processes directly involving DNA, including replication, recombination, and repair of DNA itself, expression of the genetic information via transcription, and modification of structural proteins forming the chromosome and the nuclear architecture. To date, no conclusive data are available to identify which of these processes are affected by DNA-PK activity. However, a number of proteins which possess candidate phosphate acceptor sites specific for DNA-PK and are good substrates in vitro — such as SV40 T-antigen, the generalized transcription factor Sp1, and Oct-1 — are also factors that bind to DNA and participate in transcriptional regulation, and (in the case of T-antigen) other DNA-related functions. In most of these cases the in vitro phosphorylation site(s) has not yet been shown to be phosphorylated in vivo, although preliminary evidence suggests that several sites of SV40 large T-antigen that are phosphorylated in vivo are phosphorylated by purified DNA-PK (Anderson et al. unpubl.). The functional consequences of in vitro phosphorylation of these DNA-binding substrates by DNA-PK remains to be ascertained. Clearly, such studies should be given high priority in future work.

Although every DNA sequence tested has been able to activate DNA-PK in vitro, the idea that DNA-PK is activated in vivo whenever it encounters a region of DNA free of nucleosomes is unappealing, since in that case phosphorylation of any substrate proteins in the vicinity would be indiscriminate. On the other hand, evidence suggesting that DNA-PK may bind more strongly to certain DNA sequences than to others, and the ability of DNA-PK to phosphorylate proteins that regulate transcription — in one case apparently only when the substrate protein was bound to its specific enhancer sequence — suggest a working hypothesis for specific DNA-PK function in the cell nucleus (Fig. 4). In this model, free (unbound) DNA-PK is available to bind DNA sequences which are transiently exposed to allow interactions with the molecular machinery of transcription or replication. DNA-PK binds to one or more classes of specific sequence elements (if present) and phosphorylates other proteins as they are brought into juxtaposition with DNA-PK by looping or folding of the chromatin filament, or simply by binding to the DNA at adjacent sites. This latter mechanism may well be the case for Sp1 activation, for example, which apparently requires a phosphorylation event in order to direct initiation of transcription by RNA polymerase II. Autophosphorylation of DNA-PK could eventually inactivate the enzyme, perhaps allowing it to dissociate from its specific binding site(s); in the absence of other proteins bound to the same region of DNA, inactivation might be rapid.

Phosphorylation of substrate proteins by DNA-PK could either enhance or inhibit their activity, depending upon the specific protein, the type and topography of DNA binding sequences for the two proteins, and the presence of other DNA-binding proteins which might modify the activity of DNA-PK or the ability of a particular substrate to be phosphorylated. Thus, in a circumstance such as transcription of the SV40 early region by RNA polmerase II, DNA-PK might function as a generalized transcription factor, whereas in another case (perhaps in the case of SV40 large T-antigen) DNA-PK might act to stimulate (or inhibit) the initiation of DNA replication. This general model is susceptible to direct experimental test. For example, in the case of template-dependent phosphorylation of Sp1 by DNA-PK, the model predicts that a binding site for DNA-PK should be present in a nearby sequence. The effect on Sp1 phosphorylation of modifying the relevant base sequences in synthetic oligonucleotides could be correlated with in vitro function (in this case, transcription), and the same sequences could then be tested in vivo by construction of mutant viruses or in transient expression assays. These data would complement the studies of the in vivo and in vitro phosphorylation site(s) on Sp1.

In summary, it is likely that DNA-PK plays a key role in a number of nuclear functions. Based on existing experimental data, it is now possible to construct and test models for specific effects of DNA-PK on transcription. Identification of a cognate protein in an organism susceptible to genetic manipulation, such as fission or budding yeasts, would greatly ease the task of identifying other biological functions. For the present, further elucidation of functions for DNA-PK may come from identification of candidate substrate proteins by co-immunoprecipitation and analysis of their phosphorylation sites, as well as by development of specific assays and molecular probes for studying the expression the gene encoding the enzyme and the regulation of DNA-PK activity in vivo.

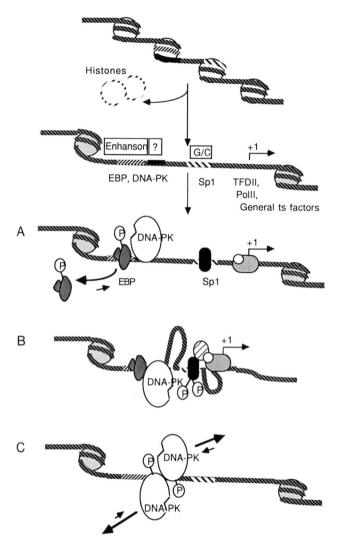

Fig. 4A-C. Model for DNA-PK role in transcription regulation. Chromatin (*top*) containing an in-dealized gene rearranges through loss of nucleosomal structure to expose enhancer elements (*boxed sites*) and promoter sequences. A putative DNA-PK binding sequence is indicated by "?", the transcription start site by "+1" and direction of transcription by the bent arrow. The transcription factor Sp1, various enhancer binding proteins (EBP), DNA-PK, general transcription factors including TFIID, and RNA polymerase II then bind to their cognate sequences. **A** diagrams a case in which DNA-PK could inhibit the activity of an EBP through phosphorylation, possibly by decreasing its affinity for DNA; **B** is the case in which DNA-PK phosphorylates a factor such as Sp1, activating it for initiation of transcription by allowing interaction with the polymerase complex, perhaps via association with a second protein (Jackson et al. 1990); **C** is the case in which DNA-PK might be inactivated by autophosphorylation (Carter et al. 1990; Lees-Miller et al. 1990), possibly decreasing its affinity for DNA

References

Arion D, Meijer L. Brizuela L, Beach D (1988) cdc2 is a component of the M phase-specific histone H1 kinase: evidence for identify with MPF. Cell 55:371–378

Baydoun H, Hoppe J, Freist W, Wagner KG (1981) Purification and ATP substrate site of a cyclic nucleotide-independent protein kinase (NI) from porcine liver nuclei J Biol Chem 257:1031–1036

Bischoff JR, Friedman PN, Marshak DR, Prives C, Beach D (1990) Human p53 in phosphorylated by p60-cdc2 and cyclin B-cdc2. Proc Nat Acad Sci USA 87:4766–4770

Carter TH, Kopman CR, James CBL (1988) DNA-stimulated protein phosphorylation in HeLa whole cell and nuclear extracts. Biochem Biophys Res Commun 157:535–540

Carter TH, Lees-Miller SP, Anderson CW (1989) Comparison of DNA-dependent protein kinases isolated from HeLa cell cytoplasmic and nuclear fractions. J Cell Biol 109(4,2):217a

Carter TH, Vancurova I, Sun I, Lou W, DeLeon S (1990) A DNA-activated protein kinase from HeLa cell nuclei. Mol Cell Biol 10 (in press)

DeCaprio JA, Ludlow JW, Lynch D, Furukawa Y, Griffin J, Piwnica-Worms H, Huang CM, Livingston DM (1989) The product of the retinoblastoma susceptibility gene has properties of a cell cycle regulatory element. Cell 58:1085–1095

Desjardins PR, Lue PF, Liew CC, Gornall AG (1972) Purification and properties of rat liver nuclear protein kinases. Can J Biochem 50:1249–1258

Dessev G, Ovocheva C, Tesheva B, Goldman R (1988) Protein kinase activity associated with the nuclear lamina. Proc Nat Acad Sci USA 85:2994–2998

Dignam JD, Lebovitz RM, Roeder RG (1983) Accurate transcription initiation by RNA polymerase II in a soluble extract from isolated mammalian nuclei. Nucl Acids Res 11:1475–1489

Draetta G, Worms HP, Morrison D, Druker B, Roberts T, Beach D (1988) Human cdc2 protein kinase is a amjor cell cycle regulated tyrosine kinase substrate. Nature (London) 336:738–744

Dufresne L, St-Pierre J, Neant I, Dube F, Guerrier P (1989) Inhibition of protein phosphorylatiaon blocks pronuclear migration and disrupts mitotic microtubules and intermediate filaments in sea urchin eggs. J Cell Biol 109:88a

Eckberg WR, Szuts EZ, Carroll AG (1987) Protein kinase C activity, protein phosphorylation and germinal vessicle breakdown in *Spisula* oocytes. Devl Biol 124:57–64

Edelman AM, Blumenthal DK, Krebs EG (1987) Protein serine/threonine kinases. Ann Rev Biochem 56:567–613

Farrell PJ, Sen GC, Dubois MF, Ratner L, Slattery E, Lengyel P (1978) Interferon action: two distinct pathways of inhibition of protein synthesis by double-stranded RNA. Proc Nat Acad Sci USA 75:5893–5897

Friedrich TD, Ingram VM (1989) Identification of a novel casein kinase activity in HeLa cell nuclei. Biochim Biophys Acta 992:41–48

Gerace L (1986) Nuclear lamina and organization of nuclear architecture. TIBS 11:443–446

Gould LK, Nurse P (1989) Tyrosine phosphorylation of the fission yeast cdc2$^+$ protein kinase regulates entry into mitosis. Nature (London) 342:39–45

Hara T, Takahashi K, Endo H (1981) Reversal of heparin inhibition of nuclear protein kinase II by polyamines and histone. FEBS Lett 128:33–39

Hashimoto E, Kuroda Y, Ku Y, Nishizuka Y (1979) Stimulation by polydeoxyribonucleotide of histone phosphorylation by guanosine 3':5'-monophosphate-dependent protein kinase. Biochem Biophys Res Commun 87:200–206

Hickey E, Brandon SE, Smale G, Lloyd D, Weber LA (1989) Sequence and regulation of a gene encoding a human 89-kilodalton heat shock protein. Mol Cell Biol 9:2615–2626

Hunt T (1989) Maturation promoting factor, cyclin and the control of M-phase. Curr Opin Cell Biol 1:268–274

Hunter T (1987) A thousand and one protein kinases. Cell 50:823–829

Hunter T, Cooper JA (1985) Protein-tyrosine kinases. Ann Rev Biochem 54:897–930

Jackson SP, MacDonald JJ, Lees-Miller S, Tjian R (1990) GC Box binding induces phosphorylation of Sp1 by a DNA-dependent protein kinase. Cell (in press)

Labbe J, Lee M, Nurse P, Picard A, Doree M (1988) Activation at M phase of a protein kinase encoded by a starfish homologue of the cell cycle control gene cdc2+. Nature (London) 335:251–254

Lamph WW, Dwarki VJ, Ofir R, Montminy M, Verma IM (1990) Negative and positive 4 regulation of transcription factor cSAMP response element-binding protein is modulated by phosphorylation. Proc Nat Acad Sci USA 87:4320–4324

Leach KL, Powers EA, Ruff VA, Jaken S, Kaufmann S (1989) Type 3 protein kinase C localization to the nuclear envelope of phorbol ester-treated NIH 3T3 cells. J Cell Biol 109:685–695

Lees-Miller SP, Anderson CW (1989a) Two human 90-kDa heat shock proteins are phosphorylated in vivo at conserved serines that are phosphorylated in vitro by casein kinase II. J Biol Chem 264:2431–2437

Lees-Miller SP, Anderson CW (1989b) The human double stranded DNA-dependent protein kinase phosphorylates the 90 kilodalton heat shock protein *hsp*90 at two amino terminal threonine residues. J Biol Chem 264:17275–17280

Lees-Miller SP, Chen YR, Anderson CW (1990) Human cells contain a DNA-activated protein kinase that phosphorylates SV40 T-antigen, p53, and other regulatory proteins. Mol Cell Biol (in press)

Lewis MK, Burgess RR (1982) Eukaryotic RNA polymerases. In: Boyer P (ed) The enzymes, Vol XV, Academic Press, NY, pp 109–145

Lund E, Paine P (1990) Non-aqueous isolation of transcriptionally active nuclei from *Xenopus* oocytes. Meth Enzymol 181:36–43

Masui Y, Markert C (1971) Cytoplasmic control of nuclear behaviour during meiotic maturation of frog oocytes. J Exp Zool 177:129–146

McVey D, Brizuela L, Mohr I, Marshak DR, Gluzman Y, Beach D (1989) Phosphorylation of large T-antigen by *cdc*2 stimulates SV40 DNA replication. Nature (Lond) 341:503–507

Meijer L, Pondaven P (1988) Cyclic activation of histone H1 kinase during sea urchin egg mitotic divisions. Exp Cell Res 174:116–129

Mimori T, Hardin JA, Steitz JA (1986) Characterization of the DNA-binding protein antigen Ku recognized by autoantibodies from patients with rheumatic disorders. J Biol Chem 261:2274–2278

Moreno S, Nurse P (1990) Substrates for p34^{cdc2}: in vivo veritas? Cell 61:549–551

Morrison DK, Kaplan DR, Escobedo JA, Rapp UR, Roberts TM, Williams LT (1989) Direct activation of the serine/threonine kinase activity of Raf-1 through tyrosine phosphorylation by the PDGF β-receptor. Cell 58:649–657

Murray AW, Kirschner MW (1989) Dominoes and clocks: the union of two views of the cell cycle. Science 246:614–621

Nigg EA, Hiltz H, Eppenberger HM, Dutly F (1985) Rapid and reversible translation of the catalytic subunit of cAMP-dependent protein kinase type II from the Golgi complex to the nucleus. EMBO J 4:2801–2806

Ohno S, Akita Y, Konno SI, Imajoh S, Suzuki K (1988) A novel phorbol ester receptor/protein kinase, nPKC, distantly related to the protein kinase C family. Cell 53:731–741

Ohtsuki K, Yamada E, Nakamura M, Ishida N (1980a) Mouse spleen nuclear protein kinases and the stimulating effect of dsDNA on NHP phosphorylation by cyclic AMP-independent protein kinase in vitro. J Biochem 87:35–45

Ohtsuki K, Shiraishi H, Yamada E, Nakamura M, Ishida N (1980b) A nonhistone chromatin protein that is a specific acceptor of nuclear cAMP-independent protein kinase from mouse spleen cells. J Biol Chem 255:2391–2395

Ohtsuki K, Shiraishi H, Sato T, Ishida N (1982) Biochemical characterization of a specific phosphate acceptor of nuclear cyclic AMP-independent protein kinase. Biochim Biophys Acta 719:32–39

Paine PL, Austerberry CF, Desjarlais LJ, Horowitz SB (1983) Protein loss during nuclear isolation. J Cell Biol 97:1240–1242

Prywes R, Dutta A, Cromlish JA, Roeder RG (1988) Phosphorylation of the serum response factor, a factor that binds to the serum response element of the c-fos enhancer. Proc Nat Acad Sci USA 85:7206–7210

Quarless SA (1985) Identification of an ionic strength sensitive nuclear protein kinase activity from the cervical carcinoma HeLa. Biochem Biophys Res Commun 133:981–987

Rose KA, Bell LE, Siefkin DA, Jacob ST (1981) A heparin-sensitive nuclear protein kinase. J Biol Chem 256:7468–7477

Sealy L, Cotton M, Chalkley R (1986) *Xenopus* nucleoplasmin: egg vs. oocyte. Biochemistry 25:3064–3072

Sen GC, Taira H, Lengyel P (1978) Interferon, double-stranded RNA, and protein phosphorylation. J Biol Chem 253:5915–5921

Sikorska M, Whitfield JF, Walker PR (1988) The regulatory and catalytic subunits of cAMP-dependent protein kinases are associated with transcriptionally active chromatin during changes in gene expression. J Biol Chem 263:3005–3011

Simpson RT (1981) Protein kinase in HeLa nucleosomes: a reevaluation of the interactions of histones with the ends of core particle DNA. Nucl Acids Res 5:1109–1119

Taylor WE, Young ET (1990) cAMP-dependent phosphorylation and inactivation of yeast transcription factor ADR1 does not affect DNA binding. Proc Nat Acad Sci USA 87:4098–4102

Thornburg W, Lindell TJ (1977) Purification of rat liver nuclear protein kinase II. J Biol Chem 252:6660–6665

Thornburg W, O'Malley AF, Lindell TJ (1978) Purification of rat liver nuclear protein kinase NI. J Biol Chem 253:4638–4641

Tuazon PT, Traugh JA (1990) Casein kinase I and II — multipotential serine protein kinases: structure, function and regulation. In: Greengard P, Robison GA (eds) Advances in second messenger and phosphoprotein research, Vol 23. (in press)

Verma R, Yu Chen K (1986) Spermine inhibits the phosphorylation of the 11,000- and 10,000-dalton nuclear proteins catalyzed by nuclear protein kinase NI in NB-15 mouse neuroblastoma cells. J Biol Chem 261:2890–2896

Walker AI, Hunt T, Jackson RJ, Anderson CW (1985) Double stranded DNA induces the phosphorylation of several proteins, including the 90,000 MW heat shock protein in animal cell extracts. EMBO J 4:139–145

Watanabe N, Vande Woude GF, Ikawa Y, Sagata N (1989) Specific proteolysis of the *c-mos* proto-oncogene product by calpain on fertilization of *Xenopus* eggs. Nature (Lond) 342:512–518

Yamamoto KK, Gonzalez GA, Biggs WH III, Montminy MR (1988) Phosphorylation-induced binding and transcriptional efficacy of nuclear factor CREB. Nature (Lond) 334:494–498

Yardeen Y, Ullrich A (1988) Growth factor receptor tyrosine kinases. Ann Rev Biochem 57:443–478

Zandomini R, Weinmann R (1984) Inhibitory effect of 5,6-dichloro-1β-D-ribofuranosylbenzimidazole on protein kinase NII. J Biol Chem 259:14804–14811

Zandomini R, Zandomini MC, Shugar D, Weinmann R (1986) Casein kinase type II is involved in the inhibition by 5,6-dichloro-1-β-D-ribofuranosylbenzimidazole of specific RNA polymerase II transcription. J Biol Chem 261:3414–3419

The Cytoskeleton During Early Development: Structural Transformation and Reorganization of RNA and Protein

D.G. CAPCO[1] and C.A. LARABELL[2]

1 Introduction

Attention has been directed toward understanding the role of the cytoskeleton in the mediation of cellular events (reviewed in Schliwa 1986). Much of this research has focused on somatic cells and has provided the foundation for theories about cytoskeletal function. Although less is known about the cytoskeletons of oocytes, zygotes, and developing embryos, it is reasonable to expect that they share common functions with the somatic cell cytoskeleton as well as exhibit some unique, specialized functions. We first summarize the function of the cytoskeleton in somatic cells, then the role of the cytoskeleton in early development.

Four distinct roles have been ascribed to the somatic cell cytoskeleton: (1) to provide structural support which maintains cell shape; (2) to facilitate intracellular translocation on organelles and other cell components; (3) to bind to RNA and proteins; and (4) to effect cell motility.

1. Structural Support. Analogous to steel beams in a building, the cytoskeleton is the infrastructure and support for the cell (reviewed in Brinkley 1981; Solomon and Magendantz 1981; Tomasek and Hay 1984; Fey et al. 1984). However, because the cell shape is not fixed like that of a building, but rather undergoes extensive changes during locomotion, progression through the cell cycle, and cytokinesis, the cytoskeleton must be capable of dynamic modifications.

2. Intracellular Translocation of Cell Components. Both microtubules and actin filaments position and translocate cellular organelles and other cytoplasmic components. As a few examples: (a) Microtubules in squid axons transport organelles between the cell body and the axon (Vale et al. 1985a,b; reviewed in Schroer and Kelley 1985; and Vale et al. 1986). Isolated squid axoplasm, or purified axon microtubules supplemented with soluble components (Vale et al. 1985) can support bidirectional movement of organelles (Koonce and Schliwa 1985; Schnapp et al. 1985). The motive force for movement away from the cell body is associated with the molecule kinesin, and movement toward the cell body appears to be associated

---

[1]Department of Zoology, Arizona State University, Tempe, AZ 85287–1501, USA
[2]Intermediate Voltage Electron Microscope Facility, Donner Laboratory, Lawrence Berkeley Laboratory, University of California, Berkeley, CA 94720, USA

with dynein (Vale et al. 1986). (b) Endosomes (vesicles formed by endocytosis) move along microtubules toward the cell center where they contact lysosomes, which are also positioned in the cell by microtubules (Herman and Albertini 1984; Matteoni and Kreis 1987). (c) Actin filaments are involved with organelle movements in cells of *Nitella*. When beads coated with heavy meromyosin are applied to the cytoplasm of ruptured *Nitella* cells, they contact the actin cables and are moved along the length of these cables (Sheetz and Spudich 1983). Similarly, organelles have been observed to move in register with actin filaments in a permeabilized freshwater amoeba (Koonce and Schliwa 1986), and myosin-coated beads move along isolated actin bundles in vitro (Sheetz and Spudich 1983; Spudich et al. 1985).

3. Binding of RNA and Proteins. A variety of components, including cellular and viral-specific mRNAs and proteins, are associated with the detergent-resistant cytoskeleton (Lenk et al. 1977; Fulton et al. 1980; Ben-Ze'ev et al. 1981; Van Venrooij et al. 1981; Ben-Ze'ev et al. 1982; Bonneau et al. 1985). Some messenger RNAs for specific cytoskeletal proteins are localized near the site where their proteins are needed. For example, actin mRNAs are concentrated at the cell periphery, a region enriched in actin filaments (Lawrence and Singer 1986). It has also been reported that protein synthesis occurs on mRNA associated with cytoskeletal components (Lenk and Penman 1979; Van Venrooij et al. 1981; Fulton and Wan 1983) and that the cellular machinery necessary for protein synthesis is associated with the cytoskeleton (Dang et al. 1983; Howe and Hershey 1984). Enzymes such as tryrosyl kinase (Cinton and Finley-Whelan 1984), tRNA synthase (Dang et al. 1983), and growth factor receptor kinase (Landreth et al. 1985) are associated with the detergent-resistant cytoskeleton. Colocalization of enzymes with the cytoskeleton has also been demonstrated using techniques other than detergent extraction. Creatine phosphokinase has been observed colocalized with intermediate filaments when examined by immunofluorescence (Eckert et al. 1980), and association of phosphofructokinase with actin filaments protects the activity of this enzyme suggesting that it is directly associated with actin in vivo (Liou and Anderson 1980; Hand and Somero 1984). Furthermore, the diffusion coefficient of fluorescently tagged aldolase, measured by fluorescence redistribution after photobleaching, suggests that aldolase is associated with a solid phase of the cytoplasm (Pagliaro and Taylor 1988), presumably the cytoskeleton. It seems clear, therefore, that both RNA and protein are in direct contact with the cytoskeleton and may depend upon cytoskeletal components to maintain their proper positions in the cell.

4. Cell Motility. Motility of somatic cells results from cytoskeletal changes in the form of contracting stress fibers, formation of adhesion plaques, and membrane flow facilitated by interaction of the cell membrane with the cytoskeleton (reviewed in Schliwa 1986; Singer and Kupfer 1986).

While a great deal remains to be determined about the role of the cytoskeleton in somatic cells, there has been an increasing interest in the role of the cytoskeleton during early development. Translocation and/or positioning of subcellular components by actin filaments or microtubules in oocytes or zygotes could be essential for providing the proper information and energy needed for the changes which occur after fertilization and may give rise to the polarity which exists in oocytes and embryos. Motility, a function that is important for the activity of the spermatozoan, is not an important cytoskeletal function in oocytes, eggs, or early embryos. However, in the late stage embryo, cell motility is important for the movement of blastomeres (giving rise to gastrulation and subsequent morphogenesis; Newport and Kirschner 1982), but the role of the cytoskeleton in later stages of development is not discussed in this review.

The studies we review support the concept that the cytoskeleton of gametes and developing cells is in many ways similar in structure and function to that of somatic cells. In addition, the studies demonstrate that some somatic cytoskeletal functions are highly amplified in oocytes and gametes and still others are unique to early development. Specialized cytoskeletal functions are to be expected since both gametes and embryos have unique roles. The formation of gametes from progenitor cells involves significant changes in both nuclear and cytoplasmic structure. Changes during oogenesis include the storage of components such as yolk, nucleic acids, proteins, and organelles, and the development of specialized organelles such as annulate lamellae and cortical granules. These modifications are accompanied by an increase in the size of the oocyte and, presumably, by a reorganization of the cytoskeleton. Upon fertilization, rapid metabolic and structural modifications occur. Changes in the levels of intracellular free calcium, the plasma membrane potential, and pH are accompanied by structural changes of the cortex following cortical granule exocytosis and formation of a mitotic or meiotic apparatus. The zygote undergoes rapid cell divisions with greatly reduced time for interphase. Many proteins, RNAs, and other components needed by the zygote are synthesized by the maternal genome and stored in the egg.

A central question for developmental biologists has been how the single cell zygote gives rise to all the tissues and organs of the organism. It has been speculated that "developmental determinants" are localized in specific parts of the oocyte and egg or that these determinants sort out during post-fertilization stages to become appropriately localized during development (Freeman 1979). These determinants are thought to bias the pathway of development in parts of the embryo, causing differences between blastomeres. These differences are then amplified by induction and cell interaction. Contemporary thought suggests that developmental determinants are composed of protein, RNA, or both. Since the cytoskeleton can act to position protein and RNA in cells, it has been suggested that the cytoskeleton plays an important role in assuring normal development by positioning the correct information at the proper location.

The unique roles of the oocyte and zygote suggest that the cytoskeleton might have specialized functions in these cells. Indeed, many eggs exhibit highly organized cytoskeletal structures containing unique components (Fig. 1A,B). As will be described in this review, specialized functions for the cytoskeleton have been identified and two of these functions appear to be ubiquitous. In all of the systems thus

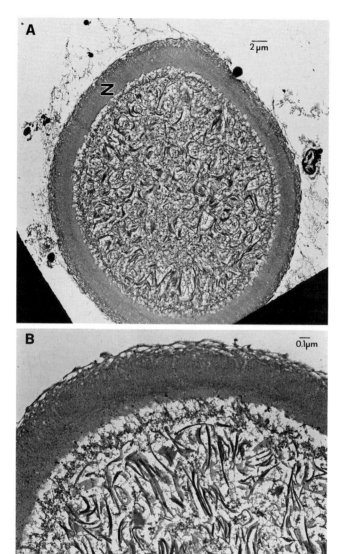

Fig. 1. A The detergent-resistant cytoskeleton of an unfertilized hamster egg. This embedment-free section shows the intricate organization in the egg, the surrounding extracellular matrix (the zona pellucida Z), and the cytoskeleton of the follicle cells. **B** Higher magnification view of the specialized cytoskeletal structures, referred to as "sheets," which are not present in somatic cells. (After Capco and McGaughey 1986)

far examined, there exists a cortical cytoskeletal domain in eggs which is reorganized during early development. In some cases the reorganization can be viewed as a contraction of the cytoskeleton after fertilization into a smaller region of the cortex, and frequently this reorganized cortical cytoskeletal domain is segregated into only one, or a select few, of the blastomeres of the embryo. The other special role of the cytoskeleton is that it binds to RNA and proteins, positioning these molecules in a strategic site within the cell.

2 The Cytoskeleton in the Early Development of Chordates

2.1 Ascidians

Pioneering studies on the role of the cytoskeleton in positioning and reorganizing RNAs during early development were conducted by Jeffery. Much of his initial work on this subject utilized marine protochordates referred to as ascidians, a classic organism of study for developmental biologists. Ascidians are commonly used because their eggs are large, easy to obtain, and transparent, allowing visualization of their differentially colored cytoplasmic components. Specific regions of the egg can be followed during embryogenesis as they become segregated into individual blastomeres and finally attain their ultimate developmental fate. In *Styela*, for example, there are three distinct cytoplasmic regions in the egg which can be followed during development. The clear ectoplasm, which is derived from germinal vesicle material that is not incorporated into the zygote nucleus, forms the ectodermal cells. The yellow-pigmented region known as myoplasm is distributed to muscle and mesenchyme lineage cells. Finally, the white yolk-filled region referred to as endoplasm forms the endodermal cells.

Developmental biologists have long speculated that cytoplasmic regions of the egg with divergent developmental potentials contain morphogenetic determinants, components which are directly responsible for the fate of that region. Such cytoplasmic determinants are thought to be composed of maternal mRNA or proteins which become sequestered in certain regions of the egg and act to bias or guide the pathway of development. If this were so, one would expect to find RNAs heterogeneously distributed in the egg and early embryo. To test this, Jeffery and his colleagues subjected histologic sections of *Styela* oocytes and embryos to in situ hybridization with nucleic acid probes to determine the spatial distribution of specific RNAs. They found that poly(A)$^+$RNA, actin mRNA, and histone mRNA have distinct spatial distributions (Jeffery and Capco 1978; Jeffery et al. 1983). Poly(A)$^+$RNA is concentrated in the germinal vesicle of the fully grown oocyte, actin mRNA is located in both the germinal vesicle, and the yellow cytoplasm found in the egg cortex, while histone mRNA is uniformly distributed throughout the cytoplasm. At fertilization, a complete reorganization of these RNAs occurs. Much of the poly(A)$^+$RNA becomes concentrated in the ectoplasm and is eventually found principally in ectodermal cells of the embryo (Jeffery et al. 1983). The yellow cytoplasm (also known as myoplasm) containing the actin mRNA becomes concentrated in the cortex of the vegetal hemisphere after fertilization and later migrates to the subequatorial region of the zygote to form the yellow crescent, the future posterior pole of the embryo. In contrast, histone mRNA maintains a uniform distribution throughout the egg cytoplasm and is uniformly partitioned into the embryonic blastomeres. The mechanisms by which these molecules are repositioned are not fully understood, but the capacity of the cytoskeleton to effect these changes was investigated by Jeffery and his colleagues.

When *Styela* eggs are detergent-extracted, the RNAs remain associated with the insoluble cytoskeleton; furthermore, the mRNAs remain in the same spatial positions occupied in unextracted eggs (Jeffery 1984). This indicates that the RNA is associated with elements of the cytoskeleton in vivo (Jeffery 1984). Further support for this concept was obtained from in situ hybridization studies conducted on eggs which had been stratified by centrifugation. During centrifugation, the different-colored cytoplasmic regions stratify into distinct bands which correspond to the ectoplasm, endoplasm, and myoplasm seen in the untreated eggs. In this case RNA remained associated with the same region with which it was identified prior to centrifugation. These studies suggest that specific RNAs which occupy distinct, localized sites in the egg are attached to the detergent-resistant cytoskeleton in that region of the egg (Jeffery 1984).

The detergent-resistant cytoskeletal network of the egg and zygote was visualized by scanning electron microscopy (Jeffery and Meier 1983, 1984). In the egg, a dense filamentous network is seen in the cortex and is referred to as the cortical cytoskeletal domain (Fig. 2). Contraction of this cortical domain occurs after fertilization, concentrating the cytoskeletal elements and associated yellow pigment in a region at the vegetal pole of the zygote which subsequently moves to a subequatorial region and forms the yellow crescent. In such zygotes, cytoskeletal filaments are very dense within the yellow crescent region, while the remainder of the embryo exhibits a much more open cytoskeletal network. Cytoskeletal filaments in the yellow crescent remain associated with detergent-resistant components of the plasma membrane (the plasma lamina) whereas the remainder of the embryo is no longer enclosed by this lamina (Jeffery and Meier 1984). The cortical contraction is accompanied by concentration of actin mRNA in the yellow crescent region as well as migration of the colored cytoplasms. Jeffery and Meier (1983) proposed that the yellow crescent became enriched in some structural component unique to this region. This idea was supported by the fact that the entire yellow crescent could be isolated after homogenization because this region remains intact while the rest of the embryo is disrupted (Jeffery 1985a); the structure which provides such integrity to this cytoplasmic region was believed to be the cytoskeleton. This information is consistent with the theory, proposed by Jeffery and Meier (1983), that a cortical cytoskeletal domain exists and that this domain contracts after fertilization. In addition, since the cytoskeletal domain moves from the vegetal pole to a subequatorial region this movement can be viewed as a segregation of the cytoskeletal domain into a region of the zygote inherited by only select blastomeres.

Sawada and Schatten (1988) demonstrated the presence of another cytoskeletal element in ascidian eggs by immunofluorescence. The entire cytoplasm of the un-

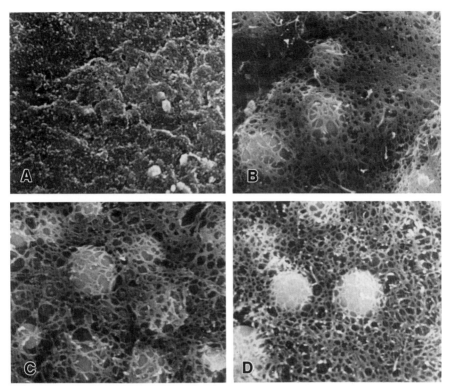

Fig. 2A-D. Scanning electron micrographs of the detergent-resistant cytoskeleton of ascidian eggs. **A** The surface of a control egg prior to detergent extraction. **B, C,** and **D** show the surface of the egg after detergent extraction and demonstrate the cortical cytoskeletal domain. Beneath the surface view of the cytoskeleton, pigment granules are observed as globular structures. (Jeffery and Meier 1983)

fertilized egg, with the exception of the germinal vesicle, is filled with a microtubule network. At fertilization, an essential reorganization occurs such that these microtubules emanate from a centrosome in the myoplasmic region at the vegetal pole. It is believed that these microtubules, in the form of an aster, play an important role in the migration of the sperm and egg pronuclei and their subsequent fusion.

Biochemical analysis of the detergent-resistant cytoskeleton demonstrates that it is composed of actin and other proteins such as intermediate filaments (Jeffery 1985a). Using a method to isolate the yellow crescent region from the remainder of the zygote, Jeffery (1985a) has shown that the yellow crescent contains approximately 15 unique polypeptides, while 43 different polypeptides are greatly enriched in the remainder of the embryo. Jeffery also isolated the mRNA from these two regions, translated it in vitro, and analyzed the protein products. The translation data showed that some mRNAs were enriched in the yellow crescent region, but none were unique to this region. There are limitations to the methods used, which may account for the absence of unique mRNAs as noted by the investigator. The mRNAs from the two regions may have mixed during the isolation procedure, the unique mRNAs may be less abundant and undetectable on the gels, unique mRNAs may synthesize neutral or basic proteins which also would be undetectable in the gel system used, unique RNAs may be untranslatable, or the phenol extraction procedure used to isolate the mRNAs may have removed proteins which could regulation in vivo translation to allow unique expression only in certain regions. These data indicate that mRNA is spatially rearranged in the zygote by a reorganization of the cytoskeleton.

In the Introduction we list four roles for the cytoskeleton of somatic cells. However, none of the somatic cell roles would have led one to expect that contraction of a cortical cytoskeletal domain would occur, nor that this segregated domain would redistribute mRNA in such an orderly fashion. This represents what is perhaps a specialized role of the cytoskeleton during early development. Such a specialized role does not rule out, however, participation of the cytoskeleton in other roles common with somatic cells, such as providing structural support or translocating organelles. Interaction of the cytoskeleton with RNA has been more clearly elucidated in ascidians than in amphibian or mammalian embryos. However, these latter two systems have provided extensive information concerning cytoskeletal reorganizations spanning more extensive time periods during development (Bement and Capco 1991; Bement et al. 1991).

2.2 Amphibians

The cytoskeleton of amphibian oocytes, eggs, and zygotes is a dynamic structure whose actin filaments, intermediate filaments, and microtubules undergo dramatic reorganizations during oogenesis and early development. These structural transitions are accompanied by a reorganization of organelles within the cell (Bement and Capco 1989a) that are each induced by a specific intracellular signal (Bement and Capco 1989b, 1990). For example, the transition of the full-grown oocyte to the meiotically mature egg includes movement of the cortical granules to a position just beneath the plasma membrane and development of an elaborate endoplasmic reticulum around them. In the subcortex, the annulate lamellae and the mitochondrial-filled corridors of cytoplasm are disrupted, while the oocyte nucleus breaks down and a meiotic spindle develops in the animal hemisphere (reviewed by Bement and Capco 1990). After fertilization, the cortical granules fuse with the plasma membrane and the contractile ring (containing actin filaments) forms during cytokinesis. It might be expected that these structural transitions are facilitated by repositioning of cytoskeletal elements.

An extensive actin filament network was visualized in the oocyte cortex (including microvilli) of three different amphibians (*Xenopus, Triturus,* and *Pleurodeles*) using electron microscopy (Franke et al. 1976). Colombo et al. (1981) and Gall et al. (1983) reported similar observations using immunofluorescent and ultrastructural examinations with anti-actin antibodies. It has been suggested that this cortical shell of actin is responsible for the cortical contraction which follows sperm penetration (Stewart-Savage and Grey 1982). In addition to the cortical network of actin, Colombo et al. (1981) reported that the yolk platelets filling the deeper cytoplasmic regions were also surrounded by actin. This actin shell was not observed by Gall et al. (1983) in his studies on pieces of the oocyte periphery which were manually isolated, detergent extracted, then viewed by either immunofluorescence or electron microscopy. It is possible, however, that the actin shell around yolk platelets could have been disrupted during isolation and extraction of the pieces. Further support for the idea of an actin shell around yolk platelets, as well as cortical granules and pigment granules, is provided by Ryabova et al. (1986). These investigators treated eggs with cytochlasin B, which disrupts actin filaments, and found that the positioning of yolk platelets, cortical granules, and pigment granules was disrupted after treatment. In addition, an elaborate cytoskeletal structure is seen around both cortical granules (Fig. 3; and Larabell and Chandler 1989) and yolk granules (Fig. 4; and Larabell, manuscript in preparation) in *Xenopus* oocytes and eggs which were quick-frozen, freeze-fractured, then deeply etched. This filamentous network makes numerous contacts with these organelles and appears to secure them to the intricate network which pervades the cytoplasm.

Two types of intermediate filaments have also been identified in amphibian eggs. Cytokeratin filaments were originally visualized beneath the cortical actin net-

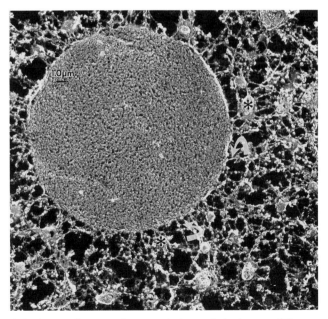

Fig. 3. Replica of the cytoskeletal network surrounding a cortical granule in the egg of the amphibian, *Xenopus laevi*. Numerous filaments are seen contacting the granule membrane (*arrows*) and its surrounding endoplasmic reticulum (*asterisks*). The specimen was fixed in glutaraldehyde, passed through distilled water, quick-frozen, freeze-fractured, deep-etched, and rotary-shadowed with platinum-carbon. This figure has been photographically reversed so that platinum deposits appear *white*

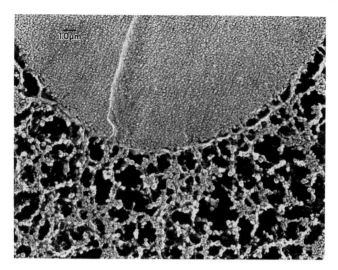

Fig. 4. Replica of the cytoskeletal network seen deeper in the egg demonstrating the numerous attachments to the membrane surrounding a yolk granule. *Xenopus laevis*. Specimen was prepared as described in Fig. 4

work by Gall et al. (1983) using immunofluorescent and immunoelectron microscopy and confirmed by Franz et al. (1983) and Godsave et al. (1984a,b). Using immunofluorescent techniques, Wylie and collaborators followed the distribution of cytokeratin from the early stages of oogenesis (pre-yolk-synthesis stages) to full-grown oocytes (post-yolk-synthesis stage). In the early oocyte, cytokeratin is observed only in the cortex and remains enriched there throughout the remainder of oogenesis. Midway through oogenesis (during yolk accumulation stages) cytokeratin can also be detected around the nucleus and appears to extend in corridors out to the cortex of the animal hemisphere, while a more punctate pattern is seen in the vegetal hemisphere. The corridors disappear during the resumption of meiotic maturation although cytokeratin filaments are still seen in the animal hemisphere (Godsave et al. 1984a). Klymkowsky et al. (1987), using immunofluorescence of intact oocytes, described a geodesic network of cytokeratin in the cortex of the vegetal hemisphere of the full-grown oocyte. However, during meiotic maturation, most of the cytokeratin network disappears (Godsave et al. 1984b; Gall and Karsenti 1987), with only very few residual cytokeratin filaments seen in the cortex of the meiotically mature egg (Klymkowsky et al. 1987). At fertilization, yet another reorganization occurs. A dense network of cytokeratin filaments develops in the vegetal hemisphere, while a sparser network of filaments is seen in the animal hemisphere. This pattern persists into the early blastula stage. Components of the sperm are not necessary for the reorganization of cytokeratin after fertilization since prick activation induces the same redistribution (Klymkowsky et al. 1987). This suggests that the development of a cytokeratin network is preprogrammed into the egg, perhaps to facilitate rapid formation of this network. Soluble cytokeratins which have been identified in oocytes and eggs of *Xenopus* (Gall and Karsenti 1987; Hauptman et al. 1989) are presumably precursors to these cytokeratin filaments (Gall and Karsenti 1987).

The distribution of another intermediate filament, vimentin, has been followed during oogenesis and meiotic maturation (Godsave et al. 1984b; Wylie et al. 1985; Tang et al. 1988). Immunofluorescent microscopy demonstrates a relatively uniform network throughout the oocyte, with a slight enrichment around the nucleus (particularly in the "mitochondrial mass") of the earliest stage (pre-yolk-synthesis) oocyte (Godsave et al. 1984b). In the full-grown oocyte, vimentin is seen in the mitochondria-filled corridors of cytoplasm in the animal hemisphere, while in the vegetal hemisphere vimentin appears in a punctate distribution. In the meiotically mature egg vimentin becomes randomly distributed throughout the egg (Godsave et al. 1984b) and remains randomized after fertilization, unlike cytokeratin (Godsave et al. 1984b; Tang et al. 1988). Thus, individual blastomeres inherit differing amounts of cytokeratin and vimentin, dependent upon the hemisphere from which the blastomere forms, and the reorganized cytoskeletal networks become differentially segregated into specific blastomeres.

Microtubules, another oocyte cytoskeletal element, also undergo extensive reorganizations during meiotic maturation and fertilization. For many years it was thought that the cytoplasm of oocytes would not support polymerization of microtubules (Heidemann and Kirschner 1975). However, several investigators have shown that a small number of microtubules do exist in the cytoplasm of oocytes (Dumont and Wallace 1972; Colman et al. 1981; Heidemann et al. 1985; Jessus et al. 1985; Jessus et al. 1986; Huchon et al. 1988). It is likely that the difficulty in detecting microtubules in the cytoplasm has been due to the large amount of cytoplasm relative to the number of microtubules. Heidemann et al. (1985) circumvented this problem by stratifying the egg cytoplasm, which concentrated the microtubules in one region and greatly facilitated their detection. It has also been suggested that microtubule-associated proteins (MAPs), which regulate the assembly of microtubules, are sequestered in oocytes (Jessus et al. 1984a; Huchon and Ozon 1985). Supporting this suggestion, Jessus et al. (1984b) have shown that MAPs can bind to free *Xenopus* ribosomes, inhibiting microtubule assembly in vitro. Microtubules may hold the nucleus in its position in the animal hemisphere, since exposure of oocytes to drugs which disrupt microtubules (colcemid and nocodazole) results in displacement of the nucleus (Lessman 1987). As the oocyte nucleus breaks down during meiotic maturation, a microtubule network becomes apparent at the vegetal side of the germinal vesicle (Huchon et al. 1981; Jessus et al. 1986). During meiotic maturation this microtubule network moves as a unit toward the animal pole and grows by assembly, but as the meiotic spindle appears, the nonmeiotic microtubule network disappears (Huchon et al. 1981).

Although the amount of polymerized tubulin decreases during meiotic maturation, microtubules are observed in the meiotically mature egg, both in the meiotic apparatus and in the form of cytoplasmic microtubules (Elinson 1983, 1985; Jessus et al. 1987; Huchon et al. 1988). At fertilization the population of microtubules is further reduced (Elinson 1985), though some microtubules are seen extending from the sperm aster. The number of microtubules increases in the second half of the first cell cycle (Elinson 1985), and an extensive array of microtubules emanate from the sperm aster to contact the female pronucleus. These microtubules appear to mediate the approach of the male and female pronuclei, at a rate of about 12 μm per minute (Steward-Savage and Grey 1982). The increased number of microtubules also coincides with the time during which the cortical cytoplasm rotates with respect to the internal cytoplasm, the time at which the dorsal/ventral axis is established. The role of microtubules in axis formation is supported by the observation that drugs which disrupt microtubules inhibit the formation of the dorsal/ventral axis in *Xenopus, Rana*, and the axolotl (Manes et al. 1978; Elinson 1983; Ubbels et al. 1983; Scharf and Gerhart 1983). In addition, Elinson and Rowning (1988) demonstrated that a highly organized array of microtubules develops in the vegetal cortex at the time of cortical rotation and may function as the "rotation motor." The modifications of the microtubule population after fertilization also occur in eggs activated without sperm, suggesting that these changes are programmed into the egg. As previously suggested, the need for this preprogramming may be due to the rapid rate at which developmental events occur after fertilization. There may be insufficient time to synthesize and position the signals required to establish the cytoskeletal networks needed in the zygote; consequently, these components may be produced and positioned by the maternal genome.

In addition to facilitating the positioning of organelles during oogenesis, meiotic maturation, and fertilization and expediting pronuclear migration and dorsal/ventral axis formation, rearrangements of the cytoskeleton may also play a role in the reorganization of RNA which occurs during these transitions.

Xenopus oocytes contain enrichments of RNAs in specific regions. In situ hybridization techniques demonstrated, for example, the localization of poly(A)$^+$RNA in the periphery of the vegetable hemisphere (Capco and Jeffery 1982; Larabell and Capco 1988). This region also contains localizations of specific mRNAs, such as tubulin mRNA (Larabell and Capco 1988) and mRNA for beta-transforming growth factor (Weeks and Melton 1987; Melton 1987). Specific RNA localizations were also demonstrated in a study in which oocytes were spatially fractionated into peripheral and central regions of each hemisphere and RNA was then purified from each region, transferred to nitrocellulose, and treated with radiolabeled probes to actin, tubulin, and histone mRNA (Perry and Capco 1988). Actin and tubulin mRNAs are concentrated in the peripheral cytoplasm of the full-grown oocyte, while histone mRNA is uniformly distributed. During meiotic maturation, tubulin mRNA becomes concentrated in the center of the egg, the concentration of actin mRNA increases slightly in the center of the egg, but histone mRNA remains uniformly distributed. After fertilization, tubulin mRNA becomes concentrated with actin mRNA in the zygote periphery and histone mRNA becomes concentrated in the animal hemisphere (Perry and Capco 1988). These and other studies indicate that meiotic maturation and fertilization are times of extensive mRNA reorganization in the egg (Capco and Jeffery 1982; Rebagliati et al. 1985; Melton 1987; Larabell and Capco 1988; Perry and Capco 1988).

The mRNA reorganizations occur coincident with a dramatic alteration of the cytoskeletal network, as described above. Only a few studies, however, have been able to link these two events. The reduction of calcium and chloride transmembrane fluxes, which occurs naturally after induction of meiotic maturation (Robinson 1979), triggers the reorganization of RNAs at meiotic maturation in vitro (Larabell and Capco 1988). The role of calcium in mediating cytoskeletal organization has been demonstrated in somatic cells (see reviews by Mooseker 1985; Schliwa 1986). It is possible, therefore, that altering calcium levels in the oocyte cytoplasm may mediate changes in cytoskeletal organization which, in turn, may initiate the RNA redistributions. We recently showed that actin and tubulin mRNA are associated with the detergent-resistant cytoskeleton, both in the full-grown oocyte and in the meiotically mature egg, providing additional support for the idea that cytoskeletal reorganizations are responsible for RNA reorganization (Hauptman et al. 1989).

The method to determine whether mRNA is associated with the cytoskeleton in *Xenopus* oocytes is based upon detergent extraction. However, the detergent-extraction approach is not easily applied to oocytes, eggs, or zygotes as large as those of *Xenopus* (1.2–1.4 mm diameter), or to multicellular embryos as large as the *Xenopus* embryo. This limitation occurs because the detergent extraction medium cannot rapidly and uniformly penetrate such a large cell or cell mass. We recently developed a method to prepare the detergent-resistant cytoskeleton of large cells or aggregates of cells (Capco et al. 1987) which we used to analyze *Xenopus* oocytes, eggs, zygotes, and embryos. Using this technique, we found that actin and tubulin mRNA are associated with the cytoskeleton and that a transition point occurs in the synthesis and assembly of cytoskeletal proteins after gastrulation. This transition point consists of two events: (1) the appearance of unique cytoskeletal proteins; and (2) an increase in the relative abundance of cytoskeletal proteins in the embryo (Hauptman et al. 1989). In these respects, the cells are becoming more like somatic cells and less like embryonic cells.

2.3 Mammals

The cytoskeletons of mammalian oocytes and embryos also play a role in oogenesis and early development. Most studies have examined the cytoskeleton in the early mammalian embryo where several unique modifications of the developmental program occur which are accompanied by dramatic alterations in the cytoskeleton and cell shape. One such modification is compaction, an event which occurs when the cleaving blastomeres of the eight- to 32-cell embryo increase their surface contact with neighboring blastomeres, forming a relatively smooth, spherical morula. At this time, blastomeres of the embryo acquire an internal/external polarity as tight junctions form between the apposed surfaces of external blastomeres. The tight junctions facilitate differential distribution of ion channels and, consequently, the movement of ions and water into the embryo interior until eventually a large, fluid-filled cavity is formed. At the blastocyst stage, the embryo is composed of the trophectoderm, an external layer of flattened cells connected by tight junctions, and the inner cell mass, a cluster of spherical blastomeres within the cavity. The cells of the trophectoderm become the extra-embryonic parts of the embryo and the inner cell mass forms mainly the embryo proper. These developmental changes are accompanied by a distinct reorganization of cytoskeleton and other cell components whose position may be mediated by the cytoskeleton.

In the Syrian hamster egg and embryo the detergent-extraction approach has revealed an extensive cytoskeletal network which contains cytoskeletal structures not previously reported in any egg or embryo (Capco and McGaughey 1986; McGaughey and Capco 1989). In addition to filamentous components which are present in both somatic cells and other types of embryos, the hamster egg contains unusual components, referred to as sheets, which are shown in Fig. 5. These planar components undergo an extensive reorganization which coincides with critical events during embryogenesis. The sheets are arranged in a whorl-like configuration throughout the internal cytoplasm of the egg but are excluded from the egg cortex. After fertilization the whorls begin to unwind, forming more linear components, and at the time of compaction these components extend throughout the blastomeres, forming foci at the portion of the plasma membrane exposed to the extracellular environment. At the blastocyst stage the cells of the inner cell mass, which will form the embryo proper, still contain sheets, but the trophectoderm cells, which will ultimately form the placenta, do not (Capco and McGaughey 1986; Fig. 5). Cytoskeletal sheets have also been identified in eggs and embryos of mice (Gallicano et al. 1991).

In addition to these unique cytoskeletal components, a cytoskeletal domain of actin filaments is seen (by immunofluorescent and ultrastructural examination) in the cortex of the hamster egg (Webster and McGaughey 1988, 1990) and the mouse egg (Ducibella et al. 1977; Lehtonen and Badley 1980). The actin domain is reorganized after fertilization and becomes polarized within each blastomere during compaction. In the mouse, the actin becomes segregated into the apical and basal regions of blastomeres and little actin is seen in the areas of cell contact that form during compaction (Ducibella et al. 1977). After compaction the actin domain undergoes further reorganization and, at the blastocyst stage, anti-actin antibodies reveal a diffuse fluorescence within the inner cell mass cells. At the same time, trophectoderm cells contain concentrations of actin filaments around their periphery (Lehtonen and Badley 1980). There is some evidence to suggest a cell-cycle dependence of this actin organization. As the mitotic apparatus forms, the amount of actin in the cortex decreases and then increases again during interphase (Ducibella et al. 1977). Myosin, which is also present in a cortical cytoskeletal domain in mammalian eggs, undergoes rearrangement during compaction and moves from the areas of contact between blastomeres to the region occupied by actin (Sobel 1983a,b, 1984).

Tubulin, unlike actin and myosin, is uniformly distributed throughout the blastomeres of the pre-compaction mouse embryo. At compaction, however, it is restricted to parallel arrays of microtubules beneath the plasma membrane in the areas of blastomere contact. The cytoplasmic microtubule array is reorganized during mitosis by disassembly and reassembly into the mitotic apparatus, but then as

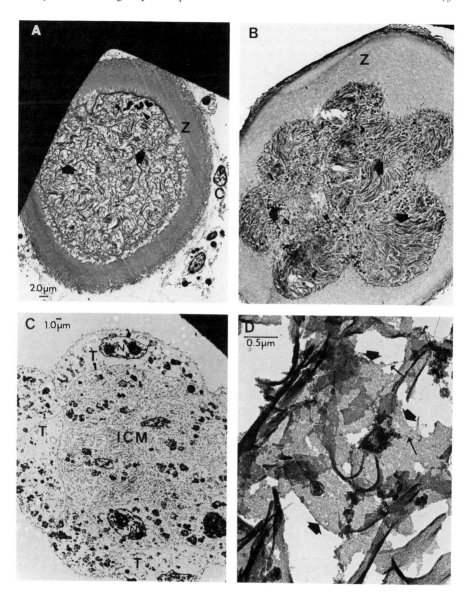

Fig. 5A-D. Series of micrographs showing the change in organization of the sheets during embryogenesis of the hamster, prepared by detergent extraction and viewed as embedment-free sections. **A** Shows the unfertilized hamster egg with the sheets in a whorled configuration (*arrows*). The *arrowhead* points to chromosomes in a forming polar body. The embryo is surrounded by the extracellular matrix, the zona pellucida (*Z*) and the follicle cells, referred to as cumulus cells (*C*). **B** At the time compaction begins, the sheets organize into long concourses (*arrows*) which extend from foci at the edge of the cell throughout the length of the blastomere. The embryo is still encased inside the zona pellucida (*Z*). **C** At the blastocyst stage the inner cell mass (*ICM*) contain sheets but the sheets are absent from the trophectoderm (*T*). Nuclei (*N*) can be seen in some of the blastomeres. **D** A higher magnification view of the sheets illustrating that they are composed of a regular array of particles (*large arrows*). In some regions they can be seen to connect with filamentous components (*small arrows*)

cytoplasmic microtubules appear during interphase the resulting newly cleaved blastomeres again contain microtubules in the areas of cell contact (Ducibella et al. 1977; Lehtonen and Badley 1980). This redistribution of microtubules coincides with the repositioning of mitochondria and lipid droplets from the areas of cell contact into a more random distribution and the microtubules may be responsible for positioning these and other cell components. At the blastocyst stage immunofluorescent examination demonstrates that tubulin is uniformly distributed in the inner cell mass, while in the trophectoderm cells tubulin appears to be concentrated in the perinuclear area (Lehtonen and Badley 1980).

The segregation of microtubules to areas of cell contact at compaction has been disputed by Houliston et al. (1987), who describe a uniform distribution of microtubules at this stage of development. This discrepancy may be explained by different experimental procedures. Houliston et al. (1987) extracted isolated groups of blastomeres in a medium containing Triton X-100. Ducibella et al. (1977) and Lehtonen and Badley (1980), who reported that microtubules were enriched in areas of cell contact, did not disassociate blastomeres into groups and did not detergent extract the embryos. We have observed that the cytoskeleton of mouse embryos extracted with Triton X-100 becomes disorganized (Mutchler et al. 1989; Gallicano et al. 1991), which could result in an artifactual rearrangement of the microtubules. In addition, further repositioning of microtubules might have occurred while the embryo was disaggregated into groups of blastomeres (i.e., doublets, triplets, and quartets) in the study by Houliston et al. 1987).

As with other organisms, pronuclear migration in the mouse egg appears to be mediated by microtubules which extend from microtubule-organizing centers and contact the pronuclei (Schatten et al. 1985, 1986a). Microtubule-organizing centers, however, appear to be derived from maternal origin, in contrast to most other systems in which the sperm provides this structure (Schatten et al. 1985, 1986a).

Intermediate filaments are also present in the early mouse embryo, although there is disagreement concerning the developmental stage at which they are detected. Jackson et al. (1981) and Paulin et al. (1980) report that intermediate filaments appear late in preimplantation development of the mouse, around the blastocyst stage, being first apparent in the outermost cells which represent presumptive trophectoderm. Lehtonen and Badley (1980), Oshima et al. (1983), and Chisholm and Houliston (1987), however, report that intermediate filaments appear as early as the two- to four-cell stage of the mouse embryo and describe a uniform distribution of intermediate filaments at this stage as well as at the blastocyst stage. In addition, Lehtonen et al. (1983) have reported the presence of intermediate filaments in the mouse oocyte. These discrepancies may result from the different techniques used. Lehtonen and Badley (1980) employed methanol or acetone in their preparations, while Oshima et al. (1983) and Lehtonen et al. (1983) employed Triton X-100 extraction. The use of solvents such as methanol or acetone may denature some proteins and expose antigenic sites which might otherwise be masked until a later stage in development. We have found (McGaughey and Capco 1989) that the hamster egg and embryo contain cytokeratin; however, the cytokeratin filaments are sequestered in the form of large "sheets" which were previously thought to be yolk.

At the time hamster embryonic cells differentiate into the trophectoderm and inner cell mass, the sheets in the trophectoderm cells disassemble, exposing 10-nm filaments. The sheets in the inner cell mass, however, do not disassemble at this developmental stage (Capco and McGaughey 1986). Similar results have been obtained for the sheets in mouse embryos (Gallicano et al. 1991). This gives the impression that cytokeratin first appears at the blastocytes stage in the trophectoderm cells, a result similar to that described by Jackson et al. (1981). However, treatment with high-salt solutions can induce disassembly of the sheets revealing cytokeratin filaments at earlier time periods (Bement et al. 1991). In addition, the use of Triton X-100 on mouse embryos severely distorts the cytoskeleton (Mutchler et al. 1989; Gallicano et al. 1991). Investigators detecting intermediate filaments at early stages of mouse embryogenesis may have employed conditions which expose intermediate filaments that are otherwise "masked".

Other cytoskeletal proteins have also been reported in mouse embryos. Alpha-actinin has been observed in a cortical cytoplasmic domain in early embryos and is enriched in trophectoderm cells during later development (Lehtonen and Badley 1980; Lehtonen et al. 1988). The role of alpha-actinin in the embryo is unknown; however, in somatic cells it is thought to represent contact points for bundling of actin filaments (Mangeat and Burridge 1983). Vinculin, a cytoskeletal protein which accumulates at sites where actin filaments appear to terminate at the plasma membrane (Geiger 1983), was found in areas of cell contact prior to compaction. Vinculin disappears after compaction and does not reappear until the blastocyst stage, where it is seen in the trophectoderm (Lehtonen et al. 1988). The membrane-associated protein, spectrin, which is frequently referred to as fodrin in cells other than red blood cells, has also been detected in early mouse embryos (Sobel and Aliegro 1985; Schatten et al. 1986b; Damjanov et al. 1986). Immunofluorescent analyses demonstrate fodrin beneath the plasma membrane of oocytes, colocalized with actin in the region of the meiotic apparatus (Schatten et al. 1986; Damjanov et al. 1986), whereas after fertilization it is enriched in cleavage furrows (Sobel and Aliegro 1985; Schatten et al. 1986b; Damjanov et al. 1986). During the remainder of preimplantation development, fodrin is reorganized into a cortical cytoskeletal domain at the periphery of each blastomere (Sobel and Aliegro 1985; Schatten et al. 1986b).

Reorganization of cytoskeletal elements during compaction is accompanied by rearrangement of cell organelles, and the synchrony with which the two events occur suggests a causal connection. Mitochondria and lipid vesicles become enriched in areas of cell contact (Ducibella et al. 1977; Wiley and Eglitis 1980), the same general area into which the microtubules segregate during interphase. In addition, the movement of these organelles is inhibited by the microtubule-destabilizing drug, colcemid (Wiley and Eglitis 1980, 1981). Further support for the idea that microtubules are involved in positioning organelles is provided from research in somatic cells, where it has been shown that microtubules serve as pathways for the transit of mitochondria and other organelles down axons (Koonce and Schliwa 1985; Vale et al. 1985, 1986a,b). Actin filaments also appear to influence the positioning of cellular organelles. Endocytotic vesicles, associated with clathrin, become sequestered in the apical region of blastomeres (Maro et al. 1985), the region where actin is located. This polarization of endocytotic vesicles can be altered by agents which disrupt actin filaments and microtubules up until the time polarity has been established (the 16-cell stage), at which point disruption of cytoskeletal elements has no effect on the positioning of endocytotic vesicles (Flemming et al. 1986).

An extensive and orderly reorganization of cytoskeletal elements occurs during mammalian development, which is similar to the restructuring which occurs in ascidian embryos due to the cortical contraction (Jeffery and Meier 1983). In the mammalian embryo the foci of sheets may also form due to a cortical contraction of the cytoskeleton within each blastomere, although this event occurs much later in development than the cortical contraction occurs in ascidians. It has been suggested that variations in the timing of developmental events often accompany evolutionary changes (Freeman 1982; Raff and Kaufman 1983). Thus, in the process of evolution from the protochordate to the mammal the timing of the cortical contraction may have changed from the time of fertilization to the time of compaction.

3 The Cytoskeleton in the Early Development of Nonchordates

3.1 Annelids

The marine annelid, *Chaetopterus*, has been used extensively by Jeffery and colleagues to map the changes in the cytoskeleton and its associated mRNA. Eggs and embryos of *Chaetopterus* exhibit cytoplasmic regions of different colors, after staining, which facilitate tracing the developmental fate of certain regions of the egg. The endoplasm contain blue-staining yolk and smaller, red-staining granules which occupy the central cytoplasmic region. The ectoplasm also contains red-staining granules which line two-thirds of the oocyte cortex, while the large oocyte nucleus (referred to as the germinal vesicle) is transparent. During maturation, fertilization, and embryogenesis these plasms migrate in a predictable pattern and partition into different blastomeres. The ectoplasm of the animal hemisphere is evenly divided between the first two blastomeres, while much of the ectoplasm in the vegetal hemisphere becomes localized in a protuberance of one of the blastomeres referred to as the polar lobe (Jeffery and Wilson 1983). Using in situ hybridization techniques, Jeffery and Wilson (1983) and Jeffery (1985b) showed that poly(A)$^+$RNA, histone mRNA, and actin mRNA are greatly enriched in the ectoplasmic region of the egg and that these RNAs migrate with the ectoplasmic region during maturation and cleavage. Jeffery showed that these mRNAs are associated with the detergent-resistant cytoskeleton in *Chaetopterus* eggs and embryos (Jeffery 1985b). He also stratified components of the living egg by centrifugation, then detergent-extracted, fixed, and processed them for in situ hybridization (Jeffery 1985b). These experiments showed that poly(A)$^+$RNA, histone mRNA, and actin mRNA are found in the centrifugal pole of the egg along with the detergent-resistant cytoskeleton. Since these mRNAs are originally in the egg cortex, and later appear at the centrifugal pole along with the cytoskeleton, it has been suggested that they are tightly associated with the cytoskeleton.

The presence of the detergent-resistant cytoskeleton appears to be requirement for normal cleavage (Swalla et al. 1985). Extended centrifugation of these eggs causes them to break into two fragments and only one of the two fragments, representing the centrifugal pole, contains the cytoskeleton. The female pronucleus, however, can be found in either the centrifugal or centripetal pole. The fragments containing the female pronucleus can be fertilized, as evidenced by the polar bodies given off at completion of meiotic maturation. Fragments containing a nucleus and cytoskeleton begin to cleave and usually develop to the blastula stage, while very few centripetal fragments, which contain the maternal nucleus but no cytoskeleton, undergo cleavage. This result suggests that the cytoskeletal domain contains factors important for normal development. Biochemical and ultrastructural examination demonstrates that the cytoskeleton-free fragments are not greatly depleted of either ribosomes or mitochondria. There are fewer yolk granules in the centripetal pole, but there is no evidence that this would inhibit the first cleavage.

One can conclude, therefore, that *Chaetopterus* eggs contain a cortical cytoskeletal domain which undergoes reorganization during development and that the presence of this cytoskeleton is essential for normal development. In addition, this cytoskeleton is associated with at least three types of mRNA [poly(A)$^+$RNA, histone mRNA, and actin mRNA] and appears to mediate the reorganization of these RNAs during development.

3.2 Oligochaetes

A cortical cytoskeletal domain has also been identified in fertilized eggs of the freshwater oligochaete, *Tubifex* (Shimizu 1984). This cortical domain is composed of actin filaments, as indicated by its binding of heavy meromyosin (Shimizu 1984) and also fluorescent rhodamine-phalloidin (Shimizu 1988). Phalloidin, which binds to filamentous actin, reveals a network of filaments at the cortex of both the animal and vegetal poles forming polar caps of actin. During the cleavage cycle, the embryo is divided into AB and CD blastomeres, then the CD blastomere further cleaves into C and D blastomeres. The developmental information for specific body parts is found in the D blastomere and its daughter cells (Shimizu 1988). The polar caps of filamentous actin are first segregated into the CD blastomere and subsequently into the D blastomere. The polar caps later fuse and become associated with some of the nuclei of the daughter cells of the D blastomere (Shimizu 1988). Organelles also become sequestered in a pattern similar to the filamentous actin (Shimizu 1982, 1986), and Shimizu (1988) speculated that the actin network may mediate this movement of organelles, in much the same way as the cortical cytoskeletal domain in ascidians is involved in the positioning of pigment (Jeffery 1985a). Treatment of the zygote with cytochalasin disrupts the cortical actin network and treatment with colchicine prevents translocation of the polar caps of filamentous actin (Shimizu 1988). Based on these results, Shimizu (1988) suggests that the microtubules of the mitotic apparatus play a role in guiding the direction of trans-location.

3.3 Nematodes

The nematode *Caenorhabditis elegans* has been studied because its embryo has a relatively small number of cells which divide in a precise, reproducible fashion, making possible a detailed fate map. With the exception of the oocyte nucleus (which is in the anterior end of the egg) there is a symmetrical distribution of all cytoplasmic components within the egg, including a cortical actin network, myosin, microtubules, yolk proteins, surface antigens, and P granules. At fertilization the sperm pronucleus becomes asymmetrically positioned in the posterior end of the zygote and the actin network becomes more pronounced and filamentous (Strome 1986; Strome and Hill 1988). Within 70–85 min after fertilization, the anterior membrane of the zygote contracts to form a temporary cleavage furrow (pseudocleavage furrow). At this time several other significant cytoplasmic events occur (Strome and Hill 1988). Much of the actin and myosin filaments segregate to the cortex at the anterior region of the zygote, although a small amount remains in the periphery of the posterior hemisphere. In addition, the P granules (developmental determinants for the germ line) segregate to the cortex of the posterior region and the female pronucleus migrates toward the posterior pole. Microtubules begin to emanate from the centrosome associated with the sperm pronucleus, which is moving anteriorly. When the sperm pronucleus and centrosome encounter the female pronucleus, the two nuclei move toward the center of the zygote, where the complex rotates. This complex later moves into the posterior region of the egg and, at cleavage, gives rise to a smaller posterior cell and a larger anterior cell. In this two-cell embryo the actin cytoskeletal domain is greatly enriched in the anterior blastomere, while the P granules are segregated into the smaller posterior blastomere. Microtubule inhibitors do not affect the P granule segregation or pseudocleavage, but they do affect pronuclear migration. Thus it appears that in this organism, as with the others already described, pronuclear migration is mediated by microtubules and there is a reorganization of a cortical actin network into one of the two blastomeres at the two-cell stage. Actin filament disruptors, such as cytochalasin D, inhibit P granule segregation and pseudo-contraction; in this case, the pronuclei meet in the center of the zygote rather than in the posterior end (Strome and Wood 1983; Hill and Strome 1988).

3.4 Insects

Karr and Alberts (1986) used immunofluorescence to show that the entire cortex of the *Drosophila* embryo contains a relatively uniform layer of actin filaments and microtubules that extends about 3 μm into the cytoplasm. An occasional microtubule-organizing center is also seen in the cortical microtubule domain. This study confirms and extends an earlier study by Warn et al. (1984), who used phalloidin to map the distribution of filamentous actin during the syncytial blastoderm stage. The early nuclear divisions in *Drosophila* occur in the center of the embryo and are not accompanied by cytokinesis. The interior nuclei are contained in islands of cytoplasm, each with a radiating network of microtubules. At cell cycle 10, the nuclei migrate to the surface and divide the cortical cytoskeletal domains of microtubules and actin into multiple cytoskeletal units around each nucleus. The microtubules which previously filled the entire cortex are reorganized (during interphase) and become concentrated at the region of the nucleus closest to the plasma membrane and appear to extend along the nucleus into the cell interior (Karr and Alberts 1986; Warn and Warn 1986). Ultrastructural analysis of an embryo two cell cycles later demonstrated similar results (Fullilove and Jacobson 1971). In addition, the cortical actin domain which previously filled the entire cortex reorganizes to form a cap-like structure between the plasma membrane and the microtubules at the apical region of the nucleus (Karr and Alberts 1986). During the subsequent nuclear divisions, the cortical cytoskeletal domain becomes further subdivided into smaller and smaller units until finally the cellular blastoderm forms and the cytoskeletal domains become permanently divided by the plasma membrane forming individual blastomeres. These results were obtained by immunofluorescence of fixed specimens, but similar results were also obtained in living cells in which fluorescently tagged anti-tubulin antibodies (Warn et al. 1987) and fluorescently tagged actin or tubulin monomers (Kellogg et al. 1988) were microinjected into embryos and their distribution traced by immunofluorescent examination. Myosin is also present in a cortical network and reorganizes into units around each nucleus as it migrates to the cell surface (Warn et al. 1980).

The above cytoskeletal reorganization coincides with the time during which the embryonic nuclei undergo developmental determination, that is between cell cycle 10 and 14. The simultaneous occurrence of these events suggests that the reorganization of the cortical cytoskeleton may be involved with the determination event.

3.5 Echinoderms

In echinoderms, as with the other organisms described, there is evidence of a cortical cytoskeletal domain in both the oocyte and the egg. Schroeder and Otto (1984) have shown, using immunofluorescence, that starfish eggs contain a cortical domain of microtubules whose appearance is cyclic and differs depending on the stage in the cell cycle. The cortical microtubule network is present during the meiotic and early mitotic divisions whenever the nuclear envelope is present (whenever the cell is in interphase), but absent whenever chromosomes are visible. Mammalian somatic cells also demonstrate cytoplasmic microtubules that are present in interphase cells, but disappear as the chromosomes condense during mitosis (Brinkley et al. 1975; Aubin et al. 1980; Brenner and Brinkley 1981). It is likely that the mechanism which controls the presence of cytoplasmic microtubules is similar in both somatic cells and these eggs. The microtubules also participate in pronuclear migration. These microtubules emanate from the centrosome of the sperm, as they do in all other organisms thus far studied except mammals, and the pronuclei appear to move along microtubules (Bestor and Schatten 1981; Schatten et al. 1985; Schatten et al. 1986a).

A cortical array of actin filaments, seen by both immunofluorescence and ultrastructural analyses, has been described in starfish eggs (Otto and Schroeder 1984) and sea urchin eggs (Spudich and Spudich 1979; Kidd and Mazia 1980; Schatten et al. 1986a; Hamaguchi and Mabuchi 1988; Henson and Begg 1988; Spudich et al. 1988). In the unfertilized sea urchin egg this actin appears to exist in two forms, a filamentous form and a nonfilamentous form (Spudich et al. 1988). Immunogold labeling demonstrated a 1-μm-thick actin shell, which encompasses the cortical granules in the egg cortex, as well as actin in the microvilli. Only the actin in the microvilli bound phalloidin, however, suggesting that the actin surrounding the cortical granules is not in a filamentous form or that the phalloidin-binding sites are masked. After fertilization a large amount of filamentous actin can be detected in the cortex, a phenomenon that has been reported in other species of echinoderms (Cline and Schatten 1986; Yonemura and Mabuchi 1987).

Fodrin is also found in a cortical cytoskeletal domain of sea urchin eggs (Schatten et al. 1986b). Prior to fertilization it is in a punctate pattern at the egg periphery, as is actin (Schatten et al. 1986b). Since fodrin is often associated with actin (Glenney and Glenney 1983; Lazarides et al. 1984), this punctate pattern may correspond to the distribution of filamentous actin seen in the microvilli (Spudich et al. 1988). After fertilization the cortical fluorescence of fodrin becomes more intense (Schatten et al. 1986b) as more filamentous actin appears near the plasma membrane. The latter event may result, in part, from the increased length of the microvilli (see review by Vacquier 1981).

Some investigators have applied the detergent-extraction approach to examine the cytoskeleton of sea urchin eggs and embryos (Moon et al. 1983). These investigators have demonstrated that more poly(A)$^+$RNA becomes associated with the cytoskeleton during embryogenesis corresponding to the increased amount of protein synthesis in the embryo. Thin section and scanning electron microscopy demonstrated the existence of a network of filaments in the cortex of the egg which increased in complexity after fertilization, corresponding to the time that RNA becomes associated with the cytoskeleton. Based on this and other work on somatic cells (Lenk and Penman 1979; Lenk et al. 1977; Van Venrooij et al. 1981), these investigators suggest that association of mRNA with the cytoskeleton may be necessary for protein synthesis.

4 Conclusion

In this chapter, we have discussed representative examples of the cytoskeletal modifications that take place during early development (from oogenesis to blastula formation) and all oocytes and embryos thus far examined have exhibited a cortical cytoskeletal domain to which two general axioms apply. First, the cytoskeleton is a well-organized cellular component which undergoes a dramatic reorganization during early development. During oogenesis and meiotic maturation of amphibian oocytes, distinct cytoskeletal modifications occur which are accompanied by a remodeling of the egg plasma membrane and rearrangement of organelles within the cells. These changes presumably are important for formation of the female pronucleus and ensuing fertilization. In eggs of many species, including nematodes, oligochaetes, amphibians, and mammals, the cytoskeleton is again modified at fertilization. In ascidian eggs, in fact, modifications at fertilization result in a significant decrease in the area of the cytoskeletal domain, an event which has been attributed to contraction of the cortical cytoskeletal network (Jeffery and Meier 1983). Whatever the mechanism for this dramatic decrease in cortical cytoskeletal area, such a reorganization has been observed in all of the systems discussed in this review. In some cases this reorganized cytoskeletal domain is eventually segregated into specific blastomeres during cleavage, an event which might affect the developmental potential of certain blastomeres in the embryo. The second general axiom about the cortical cytoskeletal domain is that the cytoskeleton in eggs of many developing systems is believed to bind RNA. We discussed experimental evidence that RNA is attached to the cytoskeleton of amphibian, annelid, ascidian, and echinoderm eggs as well as the data showing that the cytoskeleton is actually responsible for positioning RNAs in annelid and ascidian eggs.

The cortical cytoskeletal domain described in this review may be responsible for the specialized properties which developmental biologists have attributed to the egg cortex. For example, the cortex resists stratification when eggs are centrifuged at low speeds, while the internal components of many eggs are indeed stratified under these conditions (Conklin 1931; Davidson 1986). Amazingly, many eggs will develop into normal embryos after centrifugation as long as the cortical cytoplasm is not disrupted. If the speed of centrifugation is great enough to dislodge components in the egg cortex, however, abnormal development ensues, implying that information needed for normal development is securely constrained in the egg cortex. The elaborate cortical cytoskeleton described in eggs of many species may indeed provided a stabilizing matrix for those components needed in this region.

At this point, one can only speculate about the function of the cortical cytoskeletal domain. One possible function is to facilitate specialized processes that occur in the egg cortex. Oocytes, eggs, and zygotes, like other cells, receive nutrients (which must be stored) and extracellular signals (which must be transformed into intracellular signals) via plasma membrane receptors and ion channels. Mechanisms such as receptor-mediated endocytosis and signal transduction via second messengers (e.g., cyclic AMP and phosphatidylinositol pathway products) are responsible for such transfers from the outside to the inside of cells. A specialized structure such as the cytoskeleton might facilitate these processes, as suggested by the fact that the cytoskeleton is known to be involved in the stabilization of membrane receptors (Landreth et al. 1985) and in receptor-mediated endocytosis (Herman and Albertini 1984; Matteoni and Kreis 1987).

Localizations of RNA which have been identified in oocytes and eggs have, in many cases, been observed in the cortex associated with cytoskeletal elements. The role of the cortical cytoskeletal domain and associated RNA, when present, might be to mediate cellular functions and thus maintain life in this specialized cell. For example, the large size of the oocyte with a single nucleus may present special problems in maintaining metabolic activity, structural support, and intracellular signalling between the plasma membrane and the nucleus. The existence of a cortical cytoskeletal domain and localization of RNA in this region might represent a strategy to circumvent the problems generated by the large size of eggs. In addition, an elaborate cortical cytoskeleton and localized RNA might facilitate rapid responses to external cues such as those required for resumption of meiotic maturation in certain oocytes or fusion of the sperm with the egg. Sperm-egg fusion must be followed by numerous, rapid modifications to initiate the developmental program responsible for creation of a new individual. The positioning of the cytoskeleton, and consequently RNA, in the cortex may facilitate this rapid response to external stimuli.

Acknowledgements. This work was supported by NIH Grants HD23686 and HD00598.

References

Aubin JE, Osborn M, Weber K (1980) Variations in the distribution and migration of cenriole duplexes in mitotic PtK2 cells studied by immunofluorescence microscopy. J Cell Sci 43:177–194

Bement WM, Capco DG (1989a) Intracellular signals trigger ultrastructural events characteristic of meiotic maturation in *Xenopus* oocytes. Cell Tissue Res 255:183–191

Bement WM, Capco DG (1989b) Activators of protein kinase C trigger cortical granule exocytosis, cortical contraction and cleavage furrow formation in *Xenopus laevis* oocytes and eggs. J Cell Biol 108:888–892

Bement WM, Capco DG (1990a) Transformation of the amphibian oocyte into the egg: Structural and biochemical events. In: The ultrastructure of development. J Elect Microsc Tech 16:202–234

Bement WM, Capco DG (1990b) Proteine kinase C acts down stream of calciumate entry into the first mitotic interphase of *Xenopus laevis*. Cell Regulation 1:315–326

Bement WM, Capco DG (1991) Synthesis, assembly and organization of the cytoskeleton during early amphibian development. In: Seminars in Cell Biology Series: membrane skeleton in development. Saunders, New York

Bement WM, Gallicano GI, Capco DG (1991) The role of the cytoskeleton during early development. In: The ultrastructure of development. J Elect Microsc Tech (in press)

Ben-Ze'ev A, Horowitz M, Skolnik H, Abulafia R, Laub O, Aloni Y (1981) The metabolism of SV40 RNA in associated with the cytoskeletal framework. Virology 111:475–487

Ben-Ze'ev A, Abulafia R, Aloni Y (1982) SV40 virons and viral RNA metabolism are associated with cellular substructures. EMBO J 1;1225–1231

Bestor TH, Schatten G (1981) Anti-tubulin immunofluorescence microscopy of microtubules present during the pronuclear movements of sea urchin fertilization. Dev Viol 88:80–91

Bonneau A, Darveau A, Sonenberg N (1985) Effect of viral infection on host synthesis and mRNA association with the cytoplasmic cytoskeletal structure. J Cell Biol 100:1209–1218

Brenner SL, Brinkley BR (1981) Tubulin assembly sites and the organization of microtubule arrays in mammalian cells. CSHSQB 46:241–254

Brinkley BR (1981) Summary: organization of the cytoplasm. CSHSQB 46:1024–1040

Brinkley BR, Fuller GM, Highfield DP (1975) Cytoplasmic microtubules in normal and transformed cells in culture: analysis by tubulin antibody immunofluorescence. Proc Natl Acad Sci USA 72:4981–4985

Capco DG, Jeffery W (1982) Transient localizations of messenger RNA in *Xenopus laevis* oocytes. Dev Biol 89:1–12

Capco DG, McGaughey RW (1986) Cytoskeletal reorganization during early mammalian development: analysis using embedment-free sections. Dev Biol 115:446–458

Capco DG, Munoz DM, Gassman CJ (1987) A method for analysis of the detergent-resistant cytoskeleton of cells within organs. Tiss Cell 19:606–616

Chisholm JC, Houliston E (1987) Cytokeratin filament assembly in the preimplantation mouse embryo. Dev 101:565–582

Cinton GM, Finley-Whelan J (1984) Tyrosyl kinases acquired from anchorage-independent cells by a membrane-enveloped virus. J Cell Biol 99:788–795

Cline C, Schatten G (1986) Microfilaments during sea urchin fertilization: fluorescence detection with rhodaminyl phalloidin. Gamete Res 14:277–291

Colman A, Morser J, Lane C, Besley J, Wylie C, Valle G (1981) Fate of secretory proteins trapped in oocytes of *Xenopus laevis* by disruption of the cytoskeleton or by imbalanced subunit synthesis. J Cell Biol 91:770–780

Colombo R, Benedusi P, Valle G (1981) Actin in *Xenopus* development: indirect immunofluorescence study of actin localization. Differentiation 20:45–51

Conklin EG (1931) The development of centrifuged eggs of the ascidian. J Exp Zool 60:1–119

Damjanov I, Damjanov A, Lehto V-P, Virtanen I (1986) Spectrin in mouse gametogenesis and embryogenesis. Dev Biol 114:132–140

Dang CV, Yang DCH, Pollard TD (1983) Association of methionyl-tRNA synthetase with detergent-insoluble components of the rough endoplasmic reticulum. J Cell Biol 96:1138–1147

Davidson EH (1986) Gene activity in early development, 3rd edn. Academic Press, New York

Ducibella T, Ukena T, Karnovsky M, Anderson E (1977) Changes in cell surface and cortical cytoplasmic organization during early embryogenesis in the preimplantation mouse embryo. J Cell Biol 74:153–167

Dumont JN, Wallace RA (1972) The effects of vinblastine on isolated *Xenopus* oocytes. J Cell Biol 53:605–610

Eckert BS, Koons SJ, Schantz AW, Zobel CR (1980) Association of creatine phosphokinase with the cytoskeleton of cultured mammalian cells. J Cell Biol 86:1–5

Elinson R (1983) Cytoplasmic phases in the first cell cycle of the activated frog egg. Dev Biol 100:440–451

Elinson RP (1985) Changes in levels of polymeric tubulin associated with activation and dorsoventral polarization of the frog egg. Dev Biol 109:224–233

Elinson RP, Rowning B (1988) A transient array of parallel microtubules in frog eggs: potential tracks for a cytoplasmic rotation that specifies the dorsoventral axis. Dev Biol 128:185–197

Fey EG, Wan KM, Penman S (1984) Epithelial cytoskeletal framework and nuclear matrix-intermediate filament scaffold: three-dimensional organization and protein composition. J Cell Biol 98:1973–1984

Fleming TP, Cannon PM, Pickering SJ (1986) The cytoskeleton, endocytosis and cell polarity in the mouse preimplantation embryo. Dev Biol 113:406–419

Franke WW, Rathke PC, Seib E, Trendelenburg MF, Osborn M, Weber K (1976) Distribution and mode of arrangement of microfilamentous structures and actin in the cortex of the amphibian oocyte. Cytobiologie 14:111–130

Franz J, Gall L, Williams M, Picheral B, Franke W (1983) Intermediate-size filaments in a germ cell: Expression of cytokeratins in oocytes and eggs of the frog *Xenopus*. Proc Natl Acad Sci USA 80:6254–6258

Freeman G (1979) The multiple roles which cell division can play in the localization of developmental potential. In: Subtelny S, Konigsberg IR (eds) Determinants of spatial organization. Academic Press, NY, pp 53–76

Freeman G (1982) What does the comparative study of development tell us about evolution? In: Bonner JT (ed) Evolution and development. Springer, Berlin Heidelberg New York Tokyo, pp 155–167

Fullilove SL, Jacobson AG (1971) Nuclear elongation and cytokinesis in *Drosophila montana*. Develop Biol 26:560–577

Fulton AB, Wan KM (1983) Many cytoskeletal proteins associate with the HeLa cytoskeleton during translation in vitro. Cell 32:619–625

Fulton AB, Wan KM, Penman S (1980) The spatial distribution of polyribosomes in 3T3 cells and the associated assembly of proteins into the cytoskeleton. Cell 20:849–857

Gall L, Karsenti E (1987) Soluble cytokeratins in *Xenopus laevis* oocytes and eggs. Biol Cell 61:33–38

Gall L, Picheral B, Guonon P (1983) Cytochemical evidence for the presence of intermediate filaments and microfilaments in the egg of *Xenopus laevis*. Biol Cell 47:331–3422

Gallicano GI, McGaughey RW, Capco DG (1991) Cytoskeleton of the mouse egg and embryo: reorganization of planar elements. Cell Motil Cytoskel 18: 143–154

Geiger G (1983) Membrane-cytoskeleton interaction. Biochim Biophys Acta 737:305–341

Glenney JR, Glenney P (1983) Fodrin is the general spectrin-like protein found in most cells whereas spectrin and the TW protein have a restricted distribution. Cell 34:503–512

Godsave SF, Wylie CC, Lane EB, Anderton BH (1984a) Intermediate filaments in the *Xenopus* oocyte: the appearance and distribution of cytokeratin containing filaments. J Embryol Exp Morphol 83:157–167

Godsave SF, Anderton BH, Heasman J, Wylie CC (1984b) Oocytes and early embryos of *Xenopus laevis* contain intermediate filaments which react with anti-mammalian vimentin antibodies. J Embryol Exp Morphol 83:169–187

Hamaguchi Y, Mabuchi I (1988) Accumulation of fluorescently labeled actin in the cortical layer in sea urchin eggs after fertilization. Cell Motil Cytoskel 9:153–163

Hand SC, Somero GN (1984) Influence of osmolytes, thin filaments, and solubility state on elasmobranch phosphofructokinase in vitro. J Exp Zool 231:297–302

Hauptman RJ, Perry BA, Capco DG (1989) A freeze-sectioning method for preparation of the detergent-resistant cytoskeletal proteins and associated mRNA in *Xenopus* oocytes and embryos. Dev Growth Differ 31:157–164

Heidemann SR, Kirschner MW (1975) Aster formation in eggs of *Xenopus laevis*. J Cell Biol 67:105–117

Heidemann SR, Hamborg MA, Balasz JE, Lindley S (1985) Microtubules in immature oocytes of *Xenopus laevis*. J Cell Sci 77:129–141

Henson JH, Begg DA (1988) Filamentous actin organization in the unfertilized sea urchin egg cortex. Dev Biol 127:338–348

Herman B, Albertini DF (1984) A time-lapse video image intensification analysis of cytoplasmic organelle movements during endosome translocation. J Cell Biol 98:565–576

Hill DP, Strome S (1988) An analysis of the role of microfilaments in the establishment and maintenance of asymmetry in *Caenorhabditis elegans* zygotes. Dev Biol 125:75–84

Houliston E, Pickering SJ, Maro B (1987) Redistribution of microtubules and pericentriolar material during the development of polarity in mouse blastomeres. J Cell Biol 104:1299–1308

Howe JG, Hershey JWB (1984) Translational initiation factor and ribosome association with the cytoskeletal framework fraction from HeLa cells. Cell 37:85–93

Huchon D, Ozon R (1985) Microtubules during germinal vesicle breakdown (GVBD) of *Xenopus* oocytes: effect of Ca^{2+} ionophore A-23187 and taxol. Reprod Nutr Dev 25:465–479

Huchon D, Crozet N, Cantenot N, Ozon R (1981) Germinal vesicle breakdown in the *Xenopus laevis* oocyte: description of a transient microtubular structure. Reprod Nutr Dev 21:135–148

Huchon D, Jessus C, Thibier C, Ozon R (1988) Presence of microtubules in isolated cortices of prophase I and metaphase II oocytes in *Xenopus laevis*. Cell Tissue Res 254:415–420

Jackson BW, Grund C, Winter S, Franke WW, Illmensee K (1981) Formation of cytoskeletal elements during mouse embryogenesis II. Epithelial differentiation and intermediate-sized filaments in early postimplantation embryos. Differentiation 20:203–216

Jeffery WR (1984) Spatial distribution of messenger RNA in the cytoskeletal framework of ascidian eggs. Dev Biol 103:482–492

Jeffery WR (1985a) Identification of proteins and mRNAs in isolated yellow crescents of ascidian eggs. J Embryol Exp Morphol 89:275–287

Jeffery WR (1985b) The spatial distribution of maternal mRNA is determined by a cortical cytoskeletal domain in *Chaetopterus* eggs. Dev Biol 110:217–229

Jeffery WR, Capco DG (1978) Differential accumulation and localization of maternal poly(A)-containing RNA during early development of the ascidian, *Styela*. Dev Biol 67:152–166

Jeffery WR, Meier S (1983) A yellow crescent cytoskeletal domain in ascidian eggs and its role in early development. Dev Biol 96:125–143

Jeffery WR, Meier S (1984) Spatial distribution of Ooplasmic segregation of the myoplasmic actin network in stratified ascidian eggs. Roux's Arch Dev Biol 193:257–262

Jeffery WR, Wilson LJ (1983) Localization of messenger RNA in the cortex of *Chaetopterus* eggs and early embryos. J Embryol Exp Morphol 75:225–239

Jeffery WR, Tomlinson CR, Broduer RD (1983) Localization of actin mRNA during early ascidian development. Dev Biol 99:408–417

Jessus C, Friederich E, Francon J, Ozon R (1984a) In vitro inhibition of tubulin assembly by a ribonucleoprotein complex associated with the free ribosome fraction isolated from *Xenopus laevis* oocytes: effect at the level of microtubule-associated proteins. Cell Differ 14:179–187

Jessus C, Huchon D, Friederich E, Francon J, Ozon R (1984b) Interaction between rat brain microtubule associated proteins (MAPs) and free ribosomes from *Xenopus* oocyte: a possible mechanism for the in vivo distribution of MAPs. Cell Differ 14:295–301

Jessus C, Thibier C, Ozon R (1985) Identification of microtubule-associated proteins (MAPs) in *Xenopus* oocyte. Fed Eur Biochem Soc 192:135–140

Jessus C, Huchon D, Ozon R (1986) Distribution of microtubules during the breakdown of the nuclear envelope of the *Xenopus* oocyte: an immunocytochemical study. Biol Cell 56:113–120

Jessus C, Thibier C, Ozon R (1987) Levels of microtubules during the meiotic maturation of the *Xenopus* oocyte. J Cell Sci 87:705–712

Karr TL, Alberts BM (1986) Organization of the cytoskeleton in early *Drosophila* embryos. J Cell Biol 102:1494–1509

Kellogg DR, Mitchison TJ, Alberts BM (1988) Behavior of microtubules and actin filaments in living *Drosophila* embryos. Dev 103:675–686

Kidd P, Mazia D (1980) The ultrastructure of surface layers isolated from fertilized and chemically stimulated sea urchin eggs. J Ultrastruct Res 70:58–69

Klymkowsky MW, Maynell LA, Polson AG (1987) Polar asymmetry in the organization of the cortical cytokeratin system of *Xenopus laevis* oocytes and embryos. Dev 100:543–557

Koonce MP, Schliwa M (1985) Bidirectional organelle transport can occur in cell processes that contain single microtubules. J Cell Biol 100:322–326

Koonce MP, Schliwa M (1986) Reactivation of organelle movements along the cytoskeletal framework of a giant freshwater ameba. J Cell Biol 103:605–612

Landreth GE, Williams LK, Rieser GD (1985) Association of the epidermal growth factor receptor kinase with the detergent-insoluble cytoskeleton of A431 cells. J Cell Biol 101:1341–1350

Larabell CA, Capco DG (1988) Role of calcium in the localization of maternal poly(A)$^+$RNA and tubulin mRNA in *Xenopus* oocytes. Wilhelm Roux's Arch Dev Biol 197:175–183

Larabell CA, Chandler DE (1990) Quick-freeze, deep-etch, rotary-shadow views of the extracellular matrix and cortical cytoskeleton of *Xenopus laevis* eggs. J Elect Microsc Techn 13:228–243

Lawrence JB, Singer RH (1986) Intracellular localization of messenger RNAs for cytoskeletal proteins. Cell 45:407–415

Lazarides E, Nelson WJ, Kasamatsu T (1984) Segregation of two spectrin forms in the chicken optic system. Cell 36:269–278

Lehtonen E, Badley RA (1980) Localization of cytoskeletal proteins in preimplantation mouse embryos. J Embryol Exp Morph 55:211–225

Lehtonen E, Lehto VP, Vartio T, Badley RA, Virtanen I (1983) Expression of cytokeratin polypeptides in mouse oocytes and preimplantation embryos. Dev Biol 100:158–165

Lehtonen E, Ordonez G, Reima I (1988) Cytoskeleton in preimplantation mouse development. Cell Diff 24:165–178

Lenk R, Penman S (1979) The cytoskeletal framework and poliovirus metabolism. Cell 16:289–301

Lenk R, Ransom L, Kaufman Y, Penman S (1977) A cytoskeletal structure associated with polyribosomes obtained from HeLa cells. Cell 10:67–78

Lessman CA (1987) Germinal vesicle migration and dissolution in *Rana pipiens* oocytes: effect of steroids and microtubule poisons. Cell Diff 20:239–251

Liou R-S, Anderson S (1980) Activation of rabbit muscle phosphofructokinase by F-actin and reconstituted thin filaments. Biochemistry 19:2684–2688

Manes ME, Elinson RP, Barbieri FD (1978) Formation of the amphibian grey crescent; effects of colchicine and cytochalasin B. Wilhelm Roux's Arch Dev Biol 185:99–101

Mangeat PH, Burridge K (1983) Binding of HeLa spectrin to a specific HeLa membrane fraction. Cell Motil 3:657–699

Maro B, Johnson MH, Pickering SJ (1985) Changes in the distribution of membranous organelles during mouse early development. J Embryol Exp Morph 90:287–309

Matteoni R, Kreis T (1987) Translocation and clustering of endosomes and lysosomes depends on microtubules. J Cell Biol 105:1253–1265

McGaughey RW, Capco DG (1989) Specialized cytoskeletal elements in mammalian eggs: structural and biochemical evidence for their composition. Cell Motil Cytoskel 13:104–111

Melton DA (1987) Translocation of a localized maternal mRNA to the vegetal pole of *Xenopus* oocytes. Nature (Lond) 38:80–82

Moon RT, Nicosia RF, Olsen C, Hille MB, Jeffery WR (1983) The cytoskeletal framework of sea urchin eggs and embryos: developmental changes in the association of messenger RNA. Dev Biol 95:447–458

Mooseker MS (1985) Organization, chemistry, and assembly of the cytoskeletal apparatus of the intestinal brush border. In: Palade GE, Alberts BM, Spudich JA (eds) Annual review of cell biology, vol 2. Annual Reviews, Inc., Palo Alto, pp 209–241

Mutchler DV, McGaughey RW, Capco DG (1989) Cytoskeletal organization during early development in the mouse. J Cell Biol 107:605a

Newport J, Kirschner M (1982) A major developmental transition in early *Xenopus* embryos: I. Characterization and timing of cellular changes at the midblastula transition. Cell 30:657–686

Oshima RG, Howe WE, Klier FG, Adamson ED, Shevinsky LH (1983) Intermediate filament protein synthesis in preimplantation murine embryos. Dev Biol 99:447–455

Otto JJ, Schroeder TE (1984) Assembly-disassembly of actin bundles of starfish oocytes: an analysis of actin-associated proteins in the isolated cortices. Dev Biol 101:263-273

Pagliaro L, Taylor DL (1988) Aldolase exists in both the fluid and solid phases of cytoplasm. J Cell Biol 107:981–991

Paulin D, Babinet C, Weber K, Osborn M (1980) Antibodies as probes of cellular differentiation and cytoskeletal organization in the mouse blastocyst. Exp Cell Res 130:297–304

Perry BA, Capco DG (1988) Spatial reorganization of actin, tubulin and histone mRNAs during meiotic maturation and fertilization in *Xenopus* oocytes. Cell Differ Dev 25:98–108

Raff RA, Kaufmann T (1983) Embryos, genes, and evolution. In: The developmental genetic basis of evolutionary change. Macmillan, New York

Rebagliati MR, Weeks DL, Harvey RP, Melton DA (1985) Identification and cloning of localized maternal RNA's from *Xenopus* eggs. Cell 42:769–777

Robinson KR (1979) Electrical currents through full-grown and maturing *Xenopus* oocytes. Proc Natl Acad Sci USA 76:837–841

Ryabova LV, Betina MI, Vassetzky SG (1986) Influence of cytochalasin B on oocyte maturation in *Xenopus laevis*. Cell Differ 19:89–96

Sawada T, Schatten G (1988) Microtubules in ascidian eggs during meiosis, fertilization, and mitosis. Cell Motil Cytoskel 9:219–230

Scharf S, Gerhart J (1983) Axis determination in eggs of *Xenopus laevis*: a critical period before first cleavage, identified by common effects of cold, pressure and ultraviolet irradiation. Dev Biol 99:75–87

Schatten G, Simerly C, Schatten H (1985) Microtubule configurations during fertilization, mitosis, and early development in the mouse and the requirement for egg microtubule-mediated motility during mammalian fertilization. Proc Natl Acad Sci USA 82:4152–4156

Schatten H, Schatten G, Mazia D, Balczon R, Simerly C (1986a) Behavior of centrosomes during fertilization and cell division in mouse oocytes and in sea urchin eggs. Proc Natl Acad Sci USA 83:105–109

Schatten H, Cheney R, Balczon R, Willard M, Cline C, Simerly C, Schatten G (1986b) Localization of fodrin during fertilization and early development of sea urchins and mice. Dev Biol 118:457–466

Schliwa M (1986) The Cytoskeleton. An Introductory Survey. Springer, Berlin Heidelberg New York Tokyo

Schnapp BJ, Vale RD, Sheetz MP, Reese TS (1985) Single microtubules from squid axoplasm support bidirectional movement of organelles. Cell 40:455–462

Schroeder TE, Otto JJ (1984) Cyclic assembly-disassembly of cortical microtubules during maturation and early development of starfish oocytes. Dev Biol 103:493–503

Schroer TA, Kelly RB (1985) In vitro translocation of organelles along microtubules. Cell 40:729–730

Sheetz MP, Spudich JA (1983) Movement of myosin-coated fluorescent beads on actin cables in vitro. Nature (Lond) 303:31–35

Shimizu T (1982) Ooplasmic segregation in the *Tubifex* egg: mode of pole plasm accumulation and possible involvement of microfilaments. Wilhelm Roux's Arch Dev Biol 191:246–256

Shimizu T (1984) Dynamics of the actin microfilaments system in the *Tubifex* egg during ooplasmic segregation. Dev Biol 106:414–426

Shimizu T (1986) Bipolar segregation of mitochondria, actin network, and surface in the *Tubifex* egg: Role of cortical polarity. Dev Biol 116:241–251

Shimizu T (1988) Localization of actin networks during early development of *Tubifex* embryos. Dev Biol 125:321–331

Singer SJ, Kupfer A (1986) The directed migration of eukaryotic cells. In: Palade GE, Alberts BM, Spudich JA (eds) Annual review of cell biology, vol 2. Annual Reviews, Inc., Palo Alto, pp 337–365

Sobel JS (1983a) Cell-cell contact modulation of myosin organization in the early mouse embryo. Dev Biol 100:207–213

Sobel JS (1983b) Localization of myosin in the preimplantation mouse embryo. Dev Biol 95:227–231

Sobel JS (1984) Myosin rings and spreading in mouse blastomeres. J Cell Biol 99:1145–1150

Sobel JS, Alliegro MA (1985) Changes in the distribution of a spectrin-like protein during development of the preimplantation mouse embryo. J Cell Biol 100:333–336

Solomon F, Magendantz M (1981) Cytochalasin separates microtubule disassembly from loss of asymmetric morphology. J Cell Biol 89:157–161

Spudich A, Spudich JA (1979) Actin in triton-treated cortical preparations of unfertilized and fertilized sea urchin eggs. J Cell Biol 82:212–226

Spudich JA, Kron SJ, Sheetz MP (1985) Movement of myosin-coated beads on oriented filaments reconstituted from purified actin. Nature (Lond) 315:584–586

Spudich A, Wrenn JT, Wessells NK (1988) Unfertilized sea urchin eggs contain a discrete cortical shell of actin that is subdivided into two organizational states. Cell Motil Cytoskel 9:85–96

Stewart-Savege J, Grey RD (1982) The temporal and spatial relationships between cortical contraction, sperm trail formation, and pronuclear migration in fertilized *Xenopus* eggs. Wilhelm Roux's Arch Dev Biol 191:241–245

Strome S (1986) Fluorescence visualization of the distribution of microfilaments in gonads and early embryos of the nematode (*Caenorhabditis elegans*. J Cell Biol 103:2241–2252

Strome S, Hill DP (1988) Early embryogenesis in *Caenorhabditis elegans*: The cytoskeleton and spatial organization of the zygote. Bio Essays 8:145–149

Strome S, Wood WB (1983) Generation of asymmetry and segregation of germ-line granules in early *C. elegans* embryos. Cell 35:15–25

Swalla BJ, Moon RT, Jeffery WR (1985) Developmental significance of a cortical cytoskeletal domain in *Chaetopterus* eggs. Dev Biol 111:434–450

Tang P, Sharpe CR, Mohun TJ, Wylie CC (1988) Vimentin expression in oocytes, eggs and early embryos in *Xenopus laevis*. Dev 103:279–287

Tomasek JJ, Hay ED (1984) Analysis of the role of microfilaments and microtubules in acquisition of bipolarity and elongation of fibroblasts in hydrated collagen gels. J Cell Biol 99:536–549

Ubbels GA, Hara K, Koster CH, Kirschner MW (1983) Evidence for a functional role of the cytoskeleton in determination of the dorsoventral axis in *Xenopus laevis* eggs. J Embryol Exp Morphol 77:15–37

Vacquier VD (1981) Dynamic changes of the egg cortex. Dev Biol 84:1–26

Vale RD, Schnapp BJ, Reese TS, Sheetz MP (1985a) Movement of organelles along filaments dissociated from the axoplasm of the squid giant axon. Cell 40:449–454

Vale RD, Schnapp BJ, Reese TS, Sheetz MP (1985b) Organelle, beads, and microtubule translocations promoted by soluble factors from the squid giant axon. Cell 40:559–569

Vale RD, Scholey JM, Sheetz MP (1986) Kinesin: possible biological roles for a new microtubule motor. TIBS 11:464–468

Van Venrooij WJ, Sillekens PTG, Van Eekelen CAG, Reinders RJ (1981) On the association of mRNA with the cytoskeleton in uninfected and adenovirus-infected human KB cells. Exp Cell Res 135:79–91

Warn RM, Warn A (1986) Microtubule arrays present during the syncytial and cellular blastoderm stages of the early *Drosophila* embryo. Exp Cell Res 163:201–210

Warn RM, Bullard B, Magrath R (1980) Changes in the distribution of cortical myosin during the cellularization of the *Drosophila* embryo. J Embryol Exp Morphol 57:167–176

Warn RM, Magrath R, Webb S (1984) Distribution of F-actin during cleavage of the *Drosophila* syncytial blastoderm. J Cell Biol 98:156–162

Warn RM, Flegg L. Warn A (1987) An investigation of microtubule organization and functions in living *Drosophila* embryos by injection of a fluorescently labeled antibody against tyrosinated A-tubulin. J Cell Biol 105:1721–1730

Webster SD, McGaughey RW (1988) Cytoskeletal interactions between the sperm and the egg at penetration in the Syrian hamster. J Cell Biol 107:178a

Webster SD, McGaughey RW (1990) The cortical cytoskeleton and its role in sperm penetration of mammalian eggs. Dev Biol 142:61–74

Weeks DL, Melton DA (1987) A maternal mRNA localized to the vegetal hemisphere in *Xenopus* eggs codes for a growth factor related to TGF-B. Cell 51:861–867

Wiley LM, Eglitis MA (1980) Effects of colcemid on cavitation during mouse blastocoele formation. Exp Cell Res 127:89–101

Wiley LM, Eglitis MA (1981) Cell surface and cytoskeletal elements: cavitation in the mouse pre-implantation embryol. Dev Biol 86:493–501

Wylie CC, Brown D, Godsave SF, Quarmby J, Heasman J (1985) The cytoskeleton of *Xenopus* oocytes and its role in development. J Embryol Exp Morph 89:1–15

Yonemura S, Mabuchi I (1987) A wave of actin polymerization in the sea urchin egg cortex. Cell Motil Cytoskel 7:46–53

Developmental Regulations of Heat-Shock Protein Synthesis in Unstressed and Stressed Cells

O. BENSAUDE, V. MEZGER, and M. MORANGE[1]

1 Introduction

In most cells from all organisms, heat shock and several other stress promote the synthesis of proteins which have been conserved throughout evolution, the so-called heat-shock proteins (HSPs). In *Escherichia coli* a dozen HSPs have been characterized and their corresponding genes cloned and sequenced; their increased expression in response to heat shock is regulated by the accumulation of the htpR gene product (Neidhardt and Van Bogelen 1987). Under normal growth conditions the cells already express significant levels of HSPs, which are for the most part coded by essential genes. Each of the following major bacterial heat-shock genes (DnaK, GroE, HtpG) has been found to have eukaryotic multigenic homologs. The 80/90 kDa eukaryotic proteins are homologs to the bacterial htpG; the 70/80 kDa proteins are related to the bacterial dnaK; the 50/60 kDa proteins are related to the bacterial groEL. However, some eukaryotic HSPs such as ubiquitin do not appear to have a bacterial counterpart. For eukaryotes, there are confusing problems of nomenclature; HSP should designate heat-inducible proteins and HSC refer to "cognate" constitutive proteins; but the distinction between HSPs and HSCs is often a matter of semantics, constitutive or stress-inducible expression can depend either on the cell type or on different genes coding for the same protein. The letters hsp or hsc are followed by the approximate molecular weight in kDa.

HSPs and HSCs are essentially cytoplasmic proteins which may accumulate within the cell nucleus at some periods of the cell cycle or after stress. HSPs and HSCs of a given family cannot generally be separated by conventional biochemistry and are thought to play very similar roles. In eukaryotic cells proteins structurally and biochemically related to the HSPs and HSCs are found within the mitochondria, the chloroplasts, and others (GRP) are localized within the lumen of the endoplasmic reticulum. The synthesis of GRPs is stimulated by stress which affects glycosylation and by all stress which affect protein folding within the endoplasmic reticulum. All these proteins seem to display similar very general, unspecific functions of chaperoning and unfolding other proteins. Thus, HSPs, HSCs, and GRPs prevent the aggregation of nascent or unfolded polypeptides, they facilitate their assembly (or disassembly) into multiprotein structures, improve translocation through

[1]Biologie Moléculaire du Stress, Ecole Normale Supérieure, 46 rue d'Ulm, 75230 Paris Cedex 05, France

membranes, and facilitate proteolysis. We recommend the reader to seek additional information in previously published reviews (Lindquist 1986; Pelham 1986, 1989; Lindquist and Craig 1988; Ellis and Hemmingsen 1989; Rothman 1989; Horwich et al. 1990; Morimoto et al. 1990; Neupert et al. 1990).

Synthesis of heat-shock and related proteins is stress but also developmentally regulated (Bienz 1985; Bond and Schlesinger 1987). On the one hand, stress-induced proteins can be synthesized at high levels at critical stages of development in the absence of stress, on the other hand, stress induction of heat-shock protein synthesis cannot be achieved at all stages of development. In this chapter, we will attempt to illustrate some general aspects of the developmental responses, and focus on specific mammalian developmental responses.

2 Expression of Heat-Shock Genes During Gametogenesis and Early Development in the Absence of Stress

2.1 An Ancient Developmental Response: Heat-Shock Protein Hyperexpression During Sporulation and Gametogenesis

Simple eukaryotes such as yeast (*Saccharomyces*) respond to adverse environmental conditions by initiating a developmental program leading to meiosis and sporulation. Treatments which initiate sporulation are related in nature to those which trigger the heat-shock response: excess heat, solvents such as ethanol, starvation, and others. The major distinction between the developmental response and the heat-shock response may be the timing. It takes several hours for a yeast cell to sporulate and only a few minutes to mount a heat-shock response. Cyclic AMP levels seem to play a central role in sporulation of simple eukaryotes such as yeast and slime mold. Interestingly, mutant yeasts have been isolated with an altered cyclic AMP-dependent protein phosphorylation. These mutants are altered in their sporulation capacities (Matsumoto et al. 1983) and heat-shock responses (Shin et al. 1987). For example, the bcy mutants fail to synthesize the majors HSPs in response to stress and cannot enter meiosis. These mutants are deficient in the regulation of the cyclic AMP-dependent phosphorylation, the cAMP kinase is constitutively activated. Both the deficient sporulation and heat-shock response phenotype are rescued by the TPK mutation, which attenuates the catalytic properties of the cAMP kinase (Cameron et al. 1988). In contrast, the cyr1-2 mutants which produce only low levels of cyclic AMP, constitutively synthesize high levels of heat-inducible HSPs and initiate meiosis in nutrient-rich growth media. Thus, we may expect that the set of heat-shock genes overlaps the set of sporulation genes. Indeed, heat-inducible mRNAs encoding the hsp82, hsp70, and hsp26 proteins were found to increase upon yeast sporulation and the corresponding proteins accumulated (Kurtz et al. 1986; Werner-Washburne et al. 1989). A constitutive member of the 70-kDa heat-shock proteins also accumulated. The promoter of the yeast SSA3 hsp70 gene has been recently dissected and two cis-acting DNA sequences which mediate the diauxic shift response have been characterized (Boorstein and Craig 1990). These sequences are present immediately upstream and downstream of a heat-shock element which interacts positively with them. In *Dictyostelium*, a developmentally regulated membrane protein gene has also been found to be induced by heat shock (Maniak and Nellen 1988). Since the stress-induced synthesis of HSPs results in an increased resistance of the cells to stress, these proteins might also contribute to the stress resistance of spores. Expression of the yeast ubiquitin gene UBI4 is induced by stress and starvation and its presence is essential for the viability of spores and resistance to stress (Finley et al. 1987). Sporulating slime molds also accumulate mRNAs encoding ubiquitin (Giorda and Ennis 1987). The function of Hsp26 mRNA accumulation is unclear since disruption of the hsp26 gene does not affect resistance to stress, spore development, or germination (Petko and Lindquist 1986).

Heat shock and synthesis of HSPs also seem to be involved in the differentiation process of various pathogenic parasites such as the yeast/mycelia transition of the fungus *Histoplasma capsulatum* (Lambowitz et al. 1983; Caruso et al. 1987; Maresca and Kobayashi 1989); promastigote/amastigote transformation of *Leishmania* (Hunter et al. 1984; Lawrence and Robert-Gero 1985; Shapira et al. 1988; Alcina and Fresno 1988); the procyclic/trypomastigote transformation of *Tryponosoma* (Van der Ploeg 1985). Thus, early infective stages of these parasites synthesize high levels of heat-shock proteins which are found to be major antigens (Young and Elliott 1989; Young 1990).

In higher eukaryotes, meiosis and subsequent development do not depend closely on adverse environment, nevertheless, proteins homologous to the same corresponding HSP gene products accumulate during gametogenesis in *Drosophila*. Hsp26 mRNAs are found in 16-cell stage spermatocytes (Glaser et al. 1986). In the absence of heat shock, mRNAs coding for hsp83, and the 20–25 kDa HSPs are transferred from the follicular cells into the oocyte, where they accumulate (Zimmerman et al. 1983; Mason et al. 1984). As a result, hsp83 is a major protein synthesized by the unstressed developing embryo (Graziosi et al. 1980; Savoini et al. 1981). In addition, hsp70 protein encoded by the Hsc4 cognate gene is especially enriched in ovaries and embryos (Kurtz et al. 1986; Palter et al. 1986). However, neither mRNAs nor proteins corresponding to the heat-induced 70-kDa HSPs are detected. Ubiquitin mRNAs are found at high levels in *Drosophila* and *Caenorhabditis elegans* cells with no apparent developmental variations (Lee et al. 1988; Graham et al. 1989), but little is known about the corresponding polypeptide levels. It should be noted that extensive ubiquitination of histone H2A occurs in *Diptera* at the blastoderm stage (Ruder et al. 1987). Ubiquitin mRNAs are specifically found abundant in *Xenopus* unfertilized eggs and embryos during the first day of development (Dworkin-Rastl et al. 1985). A specific ubiquitin mRNA is expressed during chicken spermatogenesis (Mezquita et al. 1987). During rat and mouse spermatogenesis both the constitutive hsc70 and at least two distinct testis-specific related proteins accumulate (Anderson et al. 1982; Allen et al. 1988a,b). Testis-specific heat-shock genes are turned on at defined stages of the seminiferous epithelium (Zakeri et al. 1987; Krawczyk et al. 1988) and the corresponding mRNAs seem be translationally regulated (Zakeri et al. 1988a). One of these genes, Hsp70.2, has been cloned; it contains no introns and shows extensive similarity to the fibroblast heat-inducible Hsp70 genes within the coding region but its transcription is not stress-inducible (Zakeri et al. 1988b). Another testis-specific gene, Hsc70t, has been identified, it presents a slightly different pattern of expression (Matsumoto and Fujimoto 1990). High levels of the hsp89 proteins have also been noted in human and mouse testis (Anderson et al. 1982; Lai et al. 1984; Lee 1990). High amounts of RNAs encoding the hsp70 related grp78 protein are produced in mouse Sertoli cells (Day and Lee 1989). The corresponding protein may play an unexpected function in testes and ovaries; the C-terminal sequence of grp78 is almost identical to that of the steroidogenesis-activator polypeptide (SAP) (Li et al. 1989). SAP is a cytosolic factor which regulates mammalian steroidogenesis. The constitutive hsc70 protein is also found as a major protein synthesized by the preovulatory mouse oocyte (Curci et al. 1987) and grp94, both hsp84/86, grp78, hsc70, and hsp60 are major constituents of the matured oocyte (Fig. 1). When the oocyte matures, breakdown of the germinal vesicle results in disappearance of the Hsc70 mRNAs in the ovulated egg and the one-cell embryo after fertilization.

Is there a common pathway between stress and the developmental induction of heat-shock genes? A specific DNA sequence, the heat-shock element (HSE), is responsible for the stress-induced transcription of the heat-shock genes (Bienz and Pelham 1987). An Hsp70 *Xenopus* gene is transiently transcribed when

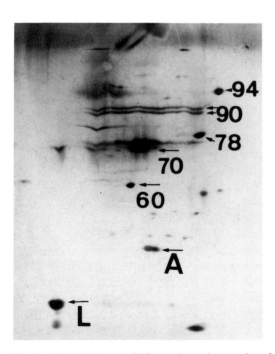

Fig. 1. HSPs, HSCs, and GRPs are the major proteins of the unfertilized mouse ovocyte. Silver staining of the proteins from 114 ovocytes separated by isoelectrophoresis followed by SDS polacrylamide gel electrophoresis. *94* grp94; *90* hsp86 and hsp84; *78* grp78; *70* hsc70; *60* hsp60; *A* actin; *L* lactate deshydrogenase

microinjected into an unstressed *Xenopus* oocyte (Bienz 1986). This constitutive expression requires both the HSE sequence and a CCAAT box; but interpretation of this observation in terms of development should take into account the fact that the corresponding endogenous gene is silent at this stage. The construction of transgenic flies has allowed the molecular dissection of the various *Drosophila* promoters. The ovarian expression of the 20–25 kDa HSPs is controlled by specific DNA sequences distinct from the HSE DNA sequences (Cohen and Meselson 1985; Hoffman et al. 1987; Xiao and Lis 1989) though some overlap may occur (Klemenz and Gehring 1986). The spermatocyte-specific expression of the Hsp26 gene is conferred by multiple regulatory elements distinct from HSEs. Two elements 5' to the transcription start conferred a correct spermatocyte expression when fused to a minimal promoter and, in addition, an element of the transcribed sequences is also able to confer a spermatocyte-specific expression (Glaser and Lis 1990).

Thus, the physiological signals which are responsible for the premeiotic expression of the heat-shock genes appear distinct from the stress signal.

2.2 Heat-Shock Proteins, First Major Products of the Zygotic Genome Transcription in Mammals

In higher eukaryotes, fertilization or parthenogenetic activation of the egg initiates development; transcription is not turned on immediately during cleavage stages, a few cell cycles occur without it (Kirschner et al. 1985; Edgar and Schubiger 1986). Nevertheless, synthesis of new proteins is observed due to translation of previously sequestered maternal mRNAs. For example, in sea urchins, hsp90 is not synthesized at the time of fertilization, but Hsp90 mRNA stored during ovogenesis becomes translated after the 64-cell stage during cleavage (Bédard and Brandhorst 1986).

In the mouse, the onset of zygotic genome transcription occurs normally at the early two-cell stage, 20 h after fertilization. The synthesis of a group of 70-kDa proteins at this stage is a major event resulting from neotranscription of the genome since it is inhibited by α-amanitin (Flach et al. 1982; Bolton et al. 1984). These proteins correspond to both the cognate hsc70 and the inducible hsp70 proteins (Bensaude et al. 1983). In vitro translation experiments demonstrate the appearance of the corresponding RNAs at the two-cell stage (Howlett and Bolton 1985).

Their synthesis does not rely on the presence of either a paternal or a maternal genome since it also occurs at the two-cell stage in both parthenogenetic and chimeric androgenetic embryos (Barra and Renard 1988). Both hsp70s appear at the right time even if replication or the first cell division is impeded by aphidicolin or X-ray irradiation (Howlett 1986; Grinfeld et al. 1987). Protein kinase(s) seem to be involved, since some inhibitors acting in vivo block the appearance of the hsp70s synthesis after the first cleavage (Poueymirou and Schultz 1989). The cAMP-dependent kinase has been proposed to play a role in the activation of transcription.

Experiments using nuclei transfer have demonstrated that a signal from the cytoplasm was involved in the onset of hsp70s synthesis at the two-cell stage (Barnes et al. 1987; Howlett et al. 1987). At the eight-cell stage, the mouse embryo does not synthesize the inducible hsp70 even after heat shock but synthesizes very high levels of the cognate hsc70. When an eight-cell stage nucleus is transferred into a one-cell stage cytoplast (devoided of its pronuclei), the reconstructed embryos do not synthesize any hsp70 immediately. A cleavage cell division is necessary, the reconstructed two-cell embryo synthesizes both the inducible hsp70 and the cognate hsc70 at the right time relative to the maturation of the recipient cytoplast; the donor nucleus genetic expression has thus been reprogrammed by the recipient cytoplast.

2.3 Constitutive Heat-Shock Protein Expression During Early Mouse Embryogenesis

Strong synthesis of proteins of the hsp70 and hsp90 family has been demonstrated in various organisms during early embryogenesis. It is spectacular at day 3 after fertilization, at the eight-cell stage in the mouse embryo; these proteins are by far the most heavily ^{35}S-methionine-labeled proteins (Morange et al. 1984). At day 8, these proteins are major constituents of the embryonic ectoderm (Bensaude and Morange 1983). In the 70-kDa range the cognate hsc70 is the major one and, in contrast with the two-cell embryo, the inducible hsp70 cannot be detected. The 90-kDa proteins display a somewhat distinct pattern of expression. Two distinct mammalian proteins belonging to the hsp90 family can be distinguished (Ullrich et al. 1986; Barnier et al. 1987). The murine hsp84 is almost identical to the human hsp89b and the murine hsp86 is almost identical to the human hsp89a (Hoffmann and Hovemann 1988; Rebbe et al. 1989; Hickey et al. 1989). In murine fibroblasts, hsp84 behaves like a cognate (high constitutive synthesis, weak stress induction) while hsp86 displays strong stress-induced and weak constitutive synthesis (Barnier et al. 1987). Both the inducible and the cognate forms are constitutively expressed at very high levels by the mouse eight-cell embryo and blastocyst. The synthesis of grp94 and grp78 is undetectable before the two-cell stage, then it increases gradually and becomes strong at the blastocyst stage (Kim et al. 1990).

2.4 Constitutive Heat-Shock Protein Expression in Mouse Embryonal Carcinoma (EC) Cells

Mouse embryos are small and afford little material for biochemical studies. Fortunately, the embryonal carcinoma (EC) stem cells have been derived as a model system which resembles the embryonic cells and can be propagated in culture. Embryonal carcinoma cells also make high levels of the same three proteins. In these cells, two distinct levels of regulation appear to occur. While the mRNAs encoding the constitutive hsc70 are rather unstable in unstressed differentiated cells such as fibroblasts, their half-life is increased in EC cells (Legagneux et al. 1989), which together with a slightly increased transcription of the gene contributes to mRNA accumulation (Giebel et al. 1988). In contrast, the Hsp86 mRNAs are very stable (Ullrich et al. 1989), and an increased transcription accounts solely for the increased expression of the protein (Legagneux et al. 1989). The grp78 promoter has been found to be very strong in F9 embryonal carcinoma cells, five- to tenfold higher than the RSV promoter and could be further stimulated threefold following treatment with the calcium ionophore A23187 (Kim et al. 1990). Embryonal carcinoma cells can be driven to differentiate in vitro upon treatment with retinoic acid and dibutyryl-cyclic AMP. Expression of 70- and 90-kDa HSPs is down-regulated along differentiation (Bensaude and Morange 1983; Levine et al. 1984; Barnier et al. 1987).

The heat-shock-induced transcription of the heat-shock genes requires the interaction of a specific transcription factor, HSF, with a specific DNA sequence, HSE (Bienz and Pelham 1987; Zimarino et al. 1990). Therefore the contribution of the heat-shock factor (HSF) to the embryonal expression was questioned since the murine HSP86 gene was both heat-inducible in fibroblasts and strongly transcribed in unstressed embryonic cells. HSF preexists in fibroblasts as a structure with no affinity for DNA; stress reveals its presence as a protein with specific affinity for heat-shock element (HSE) DNA sequences (Mezger et al. 1989). Unstressed EC cells were found to show a strong HSF activity resembling the stress-activated activity in gel retardation assays.

A different approach has been initiated by Nevins (Nevins 1982; Simon et al. 1987). The E1A adenovirus oncogene activates the transcription of both a specific human Hsp70 and the human Hsp89a genes. Since mouse EC stem cells could at least partially sustain the growth of the dl312 adenovirus deletion mutant which lacks E1A, and since the E1A-responsive adenoviral promoter, EIIa was very active in these cells, he proposed that an E1A-like cellular factor was responsible for the high level of heat-shock gene transcription. Indeed, EC stem cells contain a cellular factor which binds on the E2F sites of the EIIa promoter sequence (Reichel et al. 1987; La Thangue and Rigby 1987). This factor disappears upon in vitro differentiation with retinoic acid (La Thangue et al. 1990). The E2F factor has been originally characterized in the human HeLa cells. In differentiated cells, the E1A gene product promotes the binding of preexisting E2F cellular molecules on two sites of the EIIa promoter. The E2F-like factor from murine EC cells has slightly different properties, it does not respond to the E1A gene product and binds only one of the two E2F sites (Jansen-Durr et al. 1989; Boeuf et al. 1990). The E2F-like activity diminishes upon in vitro differentiation of the EC cells. Such a scheme might be extended to the preimplantation mouse embryos which can sustain the transcription of the EIIa gene after infection with the adenovirus dl312 deletion mutant at the eight-cell stage (Suemori et al. 1988) or sustain the transcription of the EIIa promoter coupled to various reporter gene (Dooley et al. 1989). This activity seems to disappear at about the time of implantation. The E1A gene product also activates transcription of the EIV adenovirus promoter through the binding of the E4F factor (Raychaudhuri et al. 1987). A basal EIV transcription is observed in murine EC cells which requires the same DNA sequences as E1A and c-myc transactivation (Onclercq et al. 1988). Since EC cells have high levels of c-myc protein, it has been proposed that c-myc might be the E1A-like factor.

Extensive molecular dissection of a particular human Hsp70 gene which responds to adenovirus infection has been achieved, demonstrating the role of a particular TATAT box, a CCAAT, and a purine-rich element (Wu et al. 1986; Simon et al. 1988; Williams et al. 1989). No E2F-binding sites have been characterized and no HSEs are required in mediating the E1A response. However, it should be kept in mind that the basal elements of this Hsp70 promoter function differently in human and rodent cell lines (Green et al. 1987).

2.5 High Levels of B2 Transcripts in Heat-Shocked Fibroblasts and in Undifferentiated Mouse Embryonic Cells

The HSP coding genes are transcribed by RNA polymerase II, but heat shock also strongly enhances the RNA polymerase III transcription of B2 sequences in various fibroblastic cell lines (Fornace and Mitchell 1986; Fornace et al. 1989).

Similar B2 transcripts also demonstrate a strong developmental control. B2 RNAs are present in the oocyte and increase in abundance upon fertilization (Vasseur et al. 1985; Taylor and Piko 1987). At day 4, they are abundant in the inner cell mass of the blastocyst. At day 7.5, they are present in the embryonic ectoderm and mesoderm but undetectable in the embryonic and extraembryonic endoderms. In F9 embryonal carcinoma cells, B2 sequences are also transcribed at very high levels. Upon in vitro differentiation of the F9 cells into parietal endoderm cells, the B2 gene transcription is down-regulated (White and Rigby 1989). This down-regulation appears to be due to a small decrease in RNA polymerase III activity and mainly to a reduction in transcription factor TFIIIB activity. The adenovirus E1A protein also activates RNA polymerase III transcription of B2 sequences; but the E1A activation seems to be mediated by the transcription factor TFIIIC. Thus, E1A-like factors do not seem to be involved in the developmental B2 expression.

3 Heat-Shock Protein Synthesis in Differentiation Processes

3.1 Specificities of the Heat-Shock Protein Synthesis Associated with Blood Cell Differentiation

Complex changes in the pattern of heat-shock protein synthesis have also been observed during hematopoietic differentiation. GRP78 has been characterized in pre-B lymphoid cells as an immunoglobulin heavy chain binding protein (Haas et al. 1983; Munro and Pelham 1986). The transcription of the corresponding gene is enhanced in nonsecreting B-cell myeloma lines (Nakaki et al. 1989). In chicken, Hsp70 is expressed during the maturation of the primitive erythroid lineage (Banerji et al. 1987). It peaks in day 3–4 polychromatic cells. The human erythroleukemia cell line K562 differentiates in the presence of hemin. Transcription of Hsp70 and Grp78 genes is hemin-induced in these cells (Singh and Yu 1984). Heat-shock factor binding to HSE increases with hemin treatment (Theodorakis et al. 1989). In contrast, decreased expression of hsc70 has been reported to occur as an early event during DMSO-induced differentiation of murine Friend erythroleukemia cells (Hensold and Housman 1988).

Classical heat-shock inducing agents are differentiation inducers of the HL60 human promyelocytic cell line (Yufu et al. 1989); but polar solvent-induced differentiation of this cell line is accompanied by down-regulation in hsp90 synthesis (Richards et al. 1988; Yufu et al. 1989).

3.2 Hormone-Induced Heat-Shock Protein Expression

Various differentiation processes are regulated by hormones or small diffusible substances. In this section, we will show that enhanced HSP synthesis occurs in some hormone-controlled differentiation process.

In *Drosophila*, a strong synthesis of the small heat-shock proteins (mainly hsp22, hsp23, and hsp27) but not of the other heat-shock proteins, had been noted to occur at the level of imaginal disks during pupation (Ireland et al. 1982; Sirotkin and Davidson 1982; Cheney and Shearn 1983). Pupation is regulated by the ecdysone steroid hormone. Synthesis of the same HSPs is also controlled by ecdysone in established *Drosophila* cell lines. This enhanced synthesis is in part due to an increased transcription and mRNA stability (Vitek and Berger 1984). Perturbation of chromatin architecture on ecdysone induction is found near Hsp27, Hsp23, and Hsp22 genes (Kelly and Cartwright 1989). The enhanced transcription has been analyzed by molecular dissections of the promoters using transgenic flies; the hormone responsive elements are found distinct from the heat-shock and the ovarian regulatory sequences (Mestril et al. 1986; Riddihough and Pelham 1986; Hoffman et al. 1987), though some overlap may occur (Klemenz and Gehring 1986).

Synthesis of the related hsp27 mammalian HSP but of none of the other HSPs was found to be stimulated by steroids in hamster fibroblasts (Fisher et al. 1986). The same HSP was found to be induced by estrogens in human breast cancer cell lines (Fuqua et al. 1989). The level of the same protein is associated with the differentiation state in endometrial tumors and during maturation of the cervix. Since the same HSPs are specifically induced by the same class of compounds in cells which do or do not differentiate, it is likely that the hormones are directly responsible for this induction.

The steroid-induced synthesis of the 20-kDa HSP proteins increases the thermal tolerance of the cells. This is true for *Drosophila* (Berger and Woodward 1983) and mammalian cells (Fisher et al. 1986). In mammalian cells, plasmid-driven oversynthesis of the hsp27 homolog brings per se an increased stress resistance of the corresponding cells (Landry et al. 1989). An observation which contrasts with the puzzling finding that the unique Hsp26 yeast homolog gene may be disrupted without any detectable phenotype (Susek and Lindquist 1989).

Estrogens and progestin induce the accumulation of grp94 (an endoplasmic reticulum protein of the hsp90 family) in avian oviducts both at the protein and mRNA level (Baez et al. 1987). This protein is likely to be involved in chaperoning the egg proteins secreted by the oviduct. Administration of estradiol promotes a 20-fold increase in Hsp86 mRNA, a sixfold increase in Hsp84 mRNA, and a sixfold increase in Grp94 mRNA within the uterus of ovariectomized mice (Shyamala et al. 1989). This increase is rapid. These mRNAs peak 3 h after injection and drop rapidly, but remain at levels higher than control. Administration of gonadotropin heavily stimulates the synthesis of hsp84/hsp86 in the granulosa cells which surround the maturating mouse oocytes (Amsterdam et al. 1989). This phenomena can be reproduced in vitro on cultured granulosa cells using any of the treatments which have previously been shown to bring about the maturation of granulosa cells into progesterone-producing cells; for example gonadotropins, cyclic AMP analogs, or disrupting the cytoskeleton with cytochalasin (Ben-Ze'ev and Amsterdam 1989). These responses are specific since no accumulation of the (stress-inducible) Hsp70 mRNAs or proteins was detected (data on hsp27 are lacking). The development of the mammary gland is also under the control of steroid hormones and synthesis of the 84/86 kDa HSP increases due to accumulation of the corresponding mRNAs during pregnancy and lactation in the murine mammary gland (Catelli et al. 1989).

The water mold *Achlya ambisexualis* is a filamentous fungus in which sexual reproduction is regulated by steroid hormones. The hormone treatment induced the synthesis of hsp85 (homologous to the mammalian 84/86 kDa HSPs), and this synthesis resulted from an enrichment in the corresponding mRNAs (Brunt et al. 1990).

Retinoic acid (RA) is a vitamin A derivative which seems to act as a morphogen in vertebrates. It interacts with receptors belonging to the steroid receptor superfamily. It was found to increase synthesis of the related 25-kDa HSP as well as that of the 84/86 kDa HSP in fetal mouse limb buds when pregnant mice received 100 mg/kg of RA at the time of limb bud formation (Anson et al. 1987). Retinoic acid treatment increases the thermoresistance of HeLa cells (Kim et al. 1984).

Thus, in very different systems, the hormonal regulation of heat-shock protein synthesis seems to concern always the same 20- and 90-kDa HSPs. It is noteworthy that hsp90 associates the steroid receptors in their cytoplasmic unactivated form (Catelli et al. 1985; Kost et al. 1989). This association modulates the DNA binding properties of the receptors.

3.3 Entering or Leaving a Quiescent State

The differentiation process often corresponds to changes in the cell cycle. Various reports have pointed to the variation of constitutive heat-shock gene expression during cell cycle progression.

Originally, Iida and Yahara (1984) reported that yeast, fly, or mammalian cells entering in the GO quiescent state synthesized increased amounts of various proteins related to the 70- and 90-kDa HSP family. An increase in a protein related to the cognate hsc70 has recently been demonstrated in human fibroblasts in response to serum starvation and is proposed to play a role in the increased lysosomal degradation of intracellular proteins which occurs during quiescence (Chiang et al. 1989).

Proliferation of the erythroleukemia cell K562 is suppressed by prostaglandins A1 and J2. Prostaglandins promote the synthesis of an hsp70 in this human cell line (Santoro et al. 1989). Accumulation of a 25-kDa phosphorylated protein is one of the most striking changes in the pattern of proteins synthesized by murine Ehrlich ascites tumor cells entering a stationary phase after exponential growth (Gaestel et al. 1989). This protein displays 80% identity with the human hsp27/28 protein and is likely to be the murine homolog.

Serum activation of quiescent human fibroblasts and hepatoma cells is also accompanied by an increased Hsp 70 gene and Hsp90 transcription (Ting et al. 1989). In hepatoma cells, this response was shown to be obtained with insulin but with neither platelet-derived (PDGF), epidermal growth (EGF) factors, nor phorbol ester (TPA). One particular Hsp70 serum-responsive gene has been isolated (Wu et al. 1985). Transcriptional activation of this Hsp70 gene is a rather late event which follows the burst in c-fos and c-myc expression, taking place from 8 to 12 h after addition of the serum, during the S-phase (Wu and Morimoto 1985). Population of synchronized human cells was obtained using mitotic shake. Synthesis of Hsp70 mARN peaks at the end of the S-phase, while synthesis of the corresponding proteins peaks in the M-phase (Milarski and Morimoto 1986; Celis et al. 1988). Using a distinct procedure, the Rockfeller group obtained a similar result; Hsp70 mRNAs accumulated in HeLa cells from 8 to 12 h after release from a thymidine/aphidicolin block, at the end of the S-phase (Kao et al. 1985). The corresponding HSP70 promoter harbors a serum-responsive element distinct from the HSE sequences (Wu et al. 1987). This purine-rich element might interfere with a proximal CCAAT box.

In vivo, several cells leave the quiescent state in response to an appropriate stimuli. For example, a partial hepatectomy induces cellular proliferation associated with liver regeneration. In the rat, this process is accompanied by an increased expression of heat-shock genes of the 90-kDa family and to a smaller extent of the 70-kDa family (Carr et al. 1986). For the most part mammalian T-lymphocytes exist in vivo in a quiescent or GO state. Mitogenic antibodies or lectins such as phytohemagglutinin can stimulate T-cells in vitro to enter G1 and ultimately to divide. mRNAs corresponding to the above-mentioned human Hsp70 gene accumulate upon mitogenic and IL-2 stimulation of human T-lymphocytes (Iida and Yahara 1984; Kaczmarek et al. 1987). However, the major protein to be concerned by this enhanced synthesis is the cognate hsc70. This response is not restricted solely to the 70-kDa family, but synthesis of proteins of the 90-kDa family is also enhanced in mitogen-activated murine and human T-lymphocytes (Ferris et al. 1988; Haire and O'Leary 1988; Haire et al. 1988). The enhanced HSP synthesis of mitogen-activated T-lymphocytes decreases with the age of the donor, as does the proliferative response (Faassen et al. 1989). In a more general way, the stress induction of HSP synthesis in cells decreases with age of the donor (Liu et al. 1989a,b; Fargnoli et al. 1990).

4 Deficient Heat-Shock Responses

4.1 Sporulation and Gametogenesis

Kurtz and coworkers noted that during yeast sporulation concomitantly to the increase in heat-shock-related protein expression there was a decline and finally lack of heat inducibility of one of the 70-kDa major heat-shock genes (Kurtz et al. 1986). Similar observation were made with vertebrates. Synthesis of the inducible mouse hsp70 cannot be achieved in preovulatory oocytes (Curci et al. 1987). In *Drosophila*, the stress inducibility of a construct carrying a reporter gene under control of an Hsp70 promoter could not be achieved in cells having entered meiosis (Bonner et al. 1984). In transgenic flies, late oocytes (stage 14) and maturing spermatogenic cells are the only cells which do not make the reporter protein after heat shock. In *Xenopus* oocytes, a heat-induced translation of stored Hsp70 mRNA was reported (Bienz and Gurdon 1982; Browder et al. 1987), but this observation has since been disputed and tentatively attributed to transcription in remnant follicular cells attached to the oocyte (Horrell et al. 1987; Davis and King 1989). However, heat shock clearly induces transcription of reporter genes under control of a heat-shock promoter microinjected into the oocyte (Voellmy and Rungger 1982). The different behavior of the endogenous genes might be due to the fact that microinjected DNA has not been integrated into the genome chromatin. Thus, we feel that the lack of endogenous hsp induction by stress in late oocytes or meiotic cells should be considered as a general rule which may be extended to early development and is probably due to specific changes in chromatin organization.

4.2 Early Embryogenesis

As a feature common to the developing higher eukaryotes, the stress induction of heat-shock protein synthesis cannot be achieved immediately after egg activation (Graziosi et al. 1980; Dura 1981; Roccheri et al. 1981; Bienz 1984b). However, in most species this lack of inducibility reflects the absence of transcription. In *Xenopus*, the heat-shock transcription factor is present and is heat-activable for DNA binding (Ovsenek and Heikkila 1990). As soon as the general transcription resumes, at the blastoderm stage for flies, at the blastula stage for sea urchins and toads, the major heat-shock genes become stress-inducible, but some heat-shock genes remain noninducible.

The pattern of stress-induced HSPs is developmental stage- and region-specific. For example, in *Xenopus*, induction of hsp30 does not occur before the neurula stage, while heat-inducible expression of ubiquitin, hsp70, and hsp87 is detectable earlier, in embryos which have reached the mid-blastula stage of development (Bienz 1984b; Krone and Heikkila 1988). Microinjected chimeric Hsp70/CAT and Hsp30/CAT become both heat-inducible at the mid-blastula stage (Krone and Heikkila 1989). The *Xenopus* HSP30 promoter contains the regulatory elements responsible for stress induction but the endogenous gene expression before the neural stage might depend either on repression sequences missing on the construct or on chromatin structures specific for the Hsp30 gene locus. At the neurula stage, heat-induced synthesis of a 35-kDa HSP occurs in the vegetal hemisphere cells but is not detected in the animal pole cells (Nickells and Browder 1985). This HSP has been identified as the glycolytic enzyme glyceraldehyde-3-phosphate deshydrogenase (Nickells and Browder 1988).

Mammals behave in a very specific manner. The onset of transcription in the mouse occurs on the second day after fertilization, at the two-cell stage and, as we have seen, it results in synthesis of both hsc70 and hsp70. However, on the third day while hsc70 and both the cognate and heat-inducible hsp90 are the more strongly synthesized proteins in the eight-cell mouse embryo, synthesis of the inducible hsp70 has become undetectable even after heat shock (Wittig et al. 1983; Morange et al. 1984; Muller et al. 1985; Hahnel et al. 1986). Stress does not affect the pattern of synthesized proteins. Stress inducibility of the hsp70 appears gradually during the formation of the blastocyst on the fourth day. However, stress induction of hsp105 is not detectable before implantation. All blastomeres are equivalent at the eight-cell stage, but the blastocyst contains two distinct cell populations which can be separated by surgery, the large epithelioid trophoblast cells and the small undifferentiated inner cell mass cells, both cell types being heat-inducible for hsp70 synthesis (our unpubl. data).

Similar results have been obtained with the Hsp70 promoter driven expression of a reporter enzyme. A few transgenic mice have been obtained which carry the β-galactosidase gene under control of an inducible mouse Hsp70 promoter. Expression of β-galactosidase is stress-inducible in all organs but cannot be obtained before the blastocyst stage, where it appears simultaneously in both the trophoblast and the inner cell mass (Kothary et al. 1989).

4.3 Cultured Cells

As for eight-cell stage mouse embryos, stress induction of HSP synthesis cannot be detected in a few murine EC stem cell lines such as PCC4 and PCC7-S-1009 (Wittig et al. 1983; Morange et al. 1984). The noninducible phenotype is specific to the above-mentioned EC stem cell lines. It does not occur either with other embryonal carcinoma cell lines such as F9 cells or with the embryonic stem cells (ES) which we have tested and found to be all heat-inducible for hsp70 synthesis. However, in these cell lines, as in blastocysts, induction of hsp105 synthesis was not detected (Bensaude and Morange 1983).

The lack of stress-induced HSP synthesis in EC cells is due to a lack in stress-activated transcription of the corresponding heat-shock genes (Mezger et al. 1987). Heterologous *Drosophila* Hsp70 heat-shock promoters are functional in mammalian fibroblasts and can drive the heat induction of reporter genes fused to it. However, such induction cannot be obtained when the same constructs are transfected into the noninducible PCC4 and PCC7-S-1009 EC stem cells, but in vitro differentiation promotes an inducible phenotype for the reporter gene and the endogenous inducible Hsp70 and Hsp105 genes. Since the essential features conserved between the *Drosophila* and the known mammalian promoters are the TATA box and the presence of an HSE sequence, the presence of a heat-shock element-binding factor after stress was questioned (Mezger et al. 1989). All the unstressed EC cells possess a constitutive HSE binding activity which is increased in extracts from stressed inducible EC cell such as F9. However, upon heat shock this activity is strongly depressed in extracts of noninducible EC cells such as PCC4 or PCC7-S-1009, thereby providing an interpretation for the noninducibility.

There are very few cell systems reported to be unable to mount a heat-shock response. In primitive chicken red cells, neither hsp70 synthesis nor the corresponding mRNAs are heat-shock inducible (Banerji et al. 1987). In contrast, terminal chicken red cells respond to heat shock by a strong increase in protein synthesis with little change in Hsp70 mRNAs levels. This system appears to be a very stimulating example of translational control (Banerji et al. 1984). In the human HL-60 leukemia cells, the stress-induced heat-shock protein synthesis increases after differentiation with dimethylsulfoxide and the differentiated cells appear more resistant to heat shock (Yufu et al. 1990). However, mouse plasmacytoma and Friend leukemia cell lines are defective in hsp70 synthesis after heat shock (Aujame and Morgan 1985; Aujame 1986). However, these cells, in contrast to the embryonic cells, have only a partially defective heat-shock response; synthesis of other heat-shock proteins remains stress-inducible and a stress-inducible HSE-binding activity is found in plasmacytoma cell extracts (Hensold et al. 1990). As a result, in contrast to the noniducible EC cells (our unpubl. data), these cells become thermotolerant after a priming heat shock (Aujame and Firko 1988).

5 Concluding Remarks

The stress induction of heat-shock protein synthesis results from an enhanced transcription of the heat-inducible genes, and stabilization and preferential translation of the messenger RNAs. These effects seem related to the appearance of abnormal proteins within the cells (Edington et al. 1989; Parsell and Sauer 1989). In lower eukaryotes, though stress per se can be at the origin of development, there is at present no evidence that accumulation of abnormal proteins plays a role in triggering development. In higher eukaryotes, the constitutive expressions of HSPs found in the absence of stress at some critical developmental stages does not support the idea that differentiation or development are some kind of stress. Heat-shock proteins belong to a collection of essential proteins. Stress increases the need for HSPs, probably because their chaperoning and unfolding activities prevent and cure thermal denaturation of proteins.

In addition to a protective function against an adverse environment, the extensive genome expression, genome replication and organelle biogenesis which follows activation of the egg may require the accumulation of chaperonins and unfolding proteins and may explain the partial conservation of this developmental regulation throughout evolution. Furthermore, the involvement of the HSPs and

HSCs in protein degradation, as observed in cells entering a quiescent state, might facilitate the renewal of the protein pool which accompanies some differentiation processes.

References

Alcini A, Fresno M (1988) Early and late heat-induced proteins during *Leishmania mexicana* transformation. Biochem Biophys Res Comm 156:1360–1367

Allen RL, O'Brien DA, Eddy EM (1988a) A novel hsp70-like protein (P70) is present in mouse spermatogenic cells. Mol Cell Biol 8:828–832

Allen RL, O'Brien DA, Jones CC, Rockett DL, Eddy EM (1988b) Expression of heat shock proteins by isolated mouse spermatogenic cells. Mol Cell Biol 8:3260–3266

Amsterdam A, Rotmensch S, Ben-Ze'ev A (1989) Coordinated regulation of morphological and biochemical differentiation in a steroidogenic cell: the granulosa cell model. Trends Biochem Sci 14:377–382

Anderson NL, Giometti CS, Gemmell MA, Nance SL, Anderson NG (1982) A two-dimensional electrophoretic analysis of the heat-shock-induced proteins of human cells. Clin Chem 28:1084–1092

Anson JF, Hinson WG, Pipkin JL, Kwarta RF, Hansen DK, Young JF, Burns ER, Casciano DA (1987) Retinoic acid induction of stress proteins in fetal mouse limb buds. Dev Biol 121:542–547

Ardeshir F, Flint JE, Richman SJ, Reese RT (1987) A 75 kd merozoite surface protein of *Plasmodium falciparum* which is related to the 70 kd heat-shock proteins. EMBO J 6:493–499

Arrigo AP (1987) Cellular localization of HSP23 during *Drosophila* development and following subsequent heat shock. Dev Biol 122:39–48

Arrigo AP, Pauli D (1988) Characterization of hsp27 and three immunologically related polypeptides during *Drosophila* development. Exp Cell Res 175:169–183

Aujame L (1986) Murine plasmacytomas constitute a class of natural heat-shock variants in which the major inducible hsp-68 gene is not expressed. Can J Genet Cytol 28:1064–1075

Aujame L, Firko H (1988) The major inducible heat shock protein hsp68 is not required for acquisition of thermal resistance in mouse plasmacytoma cell lines. Mol Cell Biol 8:5486–5494

Aujame L, Morgan C (1985) Nonexpression of a major heat shock gene in mouse plasmacytoma MPC-11. Mol Cell Biol 5:1780–1783

Baez M, Sargan DR, Elbrecht A, Kulomaa MS, Zarucki-Schulz T, Tsai MJ, O'Malley BW (1987) Steroid hormone regulation of the gene encoding the chicken heat shock protein Hsp 108. J Biol Chem 262:6582–6588

Banerji SS, Theodorakis NG, Morimoto RI (1984) Heat shock-induced translational control of HSP70 and globin synthesis in chicken reticulocytes. Mol Cell Biol 4:2437–2448

Banerji SS, Laing K, Morimoto RI (1987) Erythroid lineage-specific expression and inducibility of the major heat shock protein HSP70 during avian embryogenesis. Genes Dev 1:946–953

Barnes FL, Robl JM, First NL (1987) Nuclear transplantation in mouse embryos: assessment of nuclear function. Biol Reproduct 36:1267–1274

Barnier JV, Bensaude O, Monrange M, Babinet C (1987) Mouse 89 kD heat shock protein. Two polypeptides with distinct developmental regulation. Exp Cell Res 170:186–194

Barra J, Renard JP (1988) Diploid mouse embryos constructed at the late 2-cell stage from haploid parthenotes and androgenotes can develop to term. Development 102:773–779

Bédard PA, Brandhorst BP (1986) Translational activation of maternal mRNA encoding the heat-shock protein hsp90 during sea urchin embryogenesis. Dev Biol 117:286–293

Bensaude O, Babinet C, Morange M, Jacob F (1983) Heat-shock proteins, first major products of zygotic gene activity in mouse embryo. Nature (Lond) 305:331–333

Bensaude O, Morange M (1983) Spontaneous high expression of heat-shock proteins in mouse embryonal carcinoma cells and ectoderm from day 8 mouse embryo. EMBO J 2:173–177

Ben-Ze'ev A, Amsterdam A (1989) Regulation of heat shock protein synthesis by gonadotropins in cultured granulosa cells. Endocrinology 124:2584–2594

Berger E, Woodward MP (1983) Small heat shock proteins in *Drosophila* may confer thermal tolerance. Exp Cell Res 147:437–442

Bianco AE, Favalora JM, Burkot TR, Culvenor JG, Crewther PE, Brown GV, Anders RF, Coppel RL, Kemp DJ (1986) A repetitive antigen of *Plasmodium falciparum* that is homologous to heat shock protein 70 of *Drosophila melanogaster*. Proc Natl Acad Sci USA 83:8713–8717

Bienz M (1984a) *Xenopus* hsp70 genes are constitutively expressed in injected oocytes. EMBO J 3:2477–2483

Bienz M (1984b) Developmental control of the heat-shock response in *Xenopus*. Proc Natl Acad Sci USA 81:3138–3142

Bienz M (1985) Transient and developmental activation of heat-shock genes. Trends Biochem Sci 10:157–161

Bienz M (1986) A CCAAT box confers cell-type specific regulation on the *Xenopus* hsp70 gene in oocytes. Cell 46:1037–1042

Bienz M, Gurdon JB (1982) The heat-shock response in *Xenopus* oocytes is controlled at the translational level. Cell 29:811–819

Bienz M, Pelham HRB (1987) Mechanisms of heat shock gene activation in higher eukaryotes. Adv Genet 24:31–72

Boeuf H, Jansen-Durr P, Kédinger C (1990) Ela-mediated transactivation of the adenovirus Ella early promoter is restricted in undifferentiated F9 cells. Oncogene 5:691–699

Bolton VN, Oades PJ, Johnson MH (1984) The relationship between cleavage, DNA replication, and gene expression in the mouse 2-cell embryo. J Embryol Exp Morphol 79:139–163

Bond U, Schlesinger MJ (1987) Heat-shock proteins and development. Adv Genet 24:1–29

Bonner JJ, Parks C, Parker-Thornburg J, Mortin MA, Pelham HRB (1984) The use of promoter fusions in *Drosophila* genetics: isolation of mutations affecting the heat shock response. Cell 37:979–991

Boorstein WR, Craig EA (1990) Regulation of a yeast HSP70 gene by a cAMP responsive transcriptional control element. EMBO J 9:2543–2553

Borkovich KA, Farrelly FW, Finkelstein DB, Taulien J, Lindquist S (1989) hsp82 is an essential protein that is required in higher concentrations for growth of cells at higher temperatures. Mol Cell Biol 9:3919–3930

Browder LW, Heikkila JJ, Wilkes J, Wang T, Pollock M, Krone P, Ovsenek M, Kloc M (1987) Decay of the oocyte-type heat-shock response of *Xenopus laevis*. Dev Biol 124:191–199

Brunt SA, Riehl R, Silver JC (1990) Steroid hormone regulation of the *Achlya ambisexualis* 85-kilodalton heat shock protein, a component of the Achlya steroid receptor complex. Mol Cell Biol 10:273–281

Cameron S, Levin L, Zoller M, Wigler M (1988) cAMP-independent control of sporulation, glycogen metabolism, and heat shock resistance in *S. cerevisiae*. Cell 53:555–566

Carr BI, Huang TH, Buzin CH, Itakura K (1986) Induction of heat shock gene expression without heat shock by hepatocarcinogens and during hepatic regeneration in rat liver. Cancer Res 46:5106–5111

Caruso M, Sacco M, Medoff G, Maresca B (1987) Heat shock 70 gene is differentially expressed in *Histoplasma capsulatum* strains with different levels of thermotolerance and pathogenicity. Mol Microbiol 1:151–158

Catelli MG, Binart N, Jung-Testas I, Renoir JM, Baulieu EE, Feramisco JR, Welch WJ (1985) The common 90-kd protein component of non-transformed '8S' steroid receptors is a heat-shock protein. EMBO J 4:3131–3135

Catelli MG, Ramachandran C, Gauthier Y, Legagneux V, Quelard C, Baulieu EE, Shyamala G (1989) Developmental regulation of murine mammary-gland 90 kDa heat-shock proteins. Biochem J 258:895–901

Celis JE, Lauridsen JB, Basse B (1988) Cell cycle-associated change in the expression of the proliferation-sensitive and heat-shock protein hsX70 (IEF14): increased synthesis during mitosis. Exp Cell Res 177:176–185

Cheney CM, Shearn A (1983) Developmental regulation of *Drosophila* imaginal disc proteins: synthesis of a heat shock protein under non-heat-shock conditions. Dev Biol 95:325–330

Chiang HL, Terlecky SR, Plant CP, Dice JF (1989) A role for a 70-kilodalton heat shock protein in lysosomal degradation of intracellular proteins. Science 246:382–385

Cohen RS, Meselson M (1985) Separate regulatory elements for the heat-inducible and ovarian expression of the *Drosophila* hsp26 gene. Cell 43:737–746

Curci A, Bevilacqua A, Mangia F (1987) Lack of heat-shock response in preovulatory mouse oocytes. Dev Biol 123:154–160

Davis RE, King ML (1989) The developmental expression of the heat-shock response in *Xenopus laevis*. Development 105:213–222

Day AR, Lee AS (1989) Transcriptional regulation of the gene encoding the 78-kD glucose regulated protein GRP78 in mouse Sertoli cells: binding of specific factor(s) to the GRP78 promoter. DNA 301–310

Dooley TP, Miranda M, Jones NC, DePamphilis ML (1989) Transactivation of the adenovirus Ella promoter in the absence of adenovirus E1A protein is restricted to mouse oocytes and preimplantation embryos. Development 107:945–956

Dura JM (1981) Stage dependent synthesis of heat shock induced proteins isn early embryos of *Drosophila melanogaster*. Mol Gen Genet 184:381–385

Dworkin-Rastl E, Shrutkowski A, Dworkin MB (1985) Multiple ubiquitin mRNAs during *Xenopus laevis* development cotain tandem repeats of the 76 amino acid coding sequence. Cell 39:321–325

Edgar BA, Schubiger G (1986) Parameters controlling transcriptional activation during early *Drosophila* development. Cell 44:871–877

Edington BV, Whelan SA, Hightower LE (1989) Inhibition of heat shock (stress) protein induction by deuterium oxide and glycerol: additional support for the abnormal protein hypothesis of induction. J Cell Physiol 139:219–228

Ellis RJ, Hemmingsen SM (1989) Molecular chaperones: proteins essential for the biogenesis of some macromolecular structures. Trends Biochem Sci 14:339–342

Faassen AE, O'Leary JJ, Rodysill KJ, Bergh N, Hallgren HM (1989) Diminished heat-shock protein synthesis following mitogen stimulation of lymphocytes from aged donors. Exp Cell Res 183:326–334

Fargnoli J, Kusinada T, Fornace AJ, Schneider EL, Holbrook HJ (1990) Decreased expression of heat shock protein 70 mRNA and protein after heat treatment in cells of aged rats. Proc Natl Acad Sci USA 87:846–850

Ferris DK, Harel-Bellan A, Morimoto RI, Welch WJ, Farrar WL (1988) Mitogen and lymphokine stimulation of heat shock proteins in T lymphocytes. Proc Natl Acad Sci USA 85:3850–3854

Finley D, Ozkaynak E, Varshavsky A (1987) The yeast polyubiquitin gene is essential for resistance to high temperatures, starvation and other stresses. Cell 48:1035–1046

Fisher GA, Anderson RL, Hahn GM (1986) Glucocorticoid-induced heat resistance in mammalian cells. J Cell Physiol 128:127–132

Flach G, Johnson MH, Braude PR, Taylor RAS, Bolton VN (1982) The transition from maternal to embryonic control in the 2-cell mouse embryo. EMBO J 1:681–686

Fornace AJ, Alamo I, Hollander MC (1988) DNA damage-inducible transcripts in mammalian cells. Proc Natl Acad Sci USA 85:8800–8804

Fornace AJ, Alamo I, Hollander MC, Lamoreaux E (1989) Induction of heat shock protein transcripts and B2 transcripts by various stresses in Chinese hamster cells. Exp Cell Res 182:61–74

Fornace AJ, Mitchell JB (1986) Induction of B2 RNA polymerase III transcription by heat shock: enrichment for heat shock induced sequences in rodent cells by hybridization subtraction. Nucl Acid Res 14:5793–5811

Fuqua SAW, Blum-Salingaros M, McGuire WL (1989) Induction of the estrogen-regulated "24K" protein by heat shock. Cancer Res 49:4126–4129

Gaestel M, Gross B, Benndorf R, Strauss M, Schunk WH, Kraft R, Otto A, Böhm H, Stahl J, Drabsch H, Bielka H (1989) Molecular cloning, sequencing and expression in *Escherichia coli* of the 24-kDa growth-related protein of Ehrlich ascites tumor and its homology to mammalian stress proteins. Eur J Biochem 179:209–213

Giebel LB, Dworniczak BP, Bautz EKF (1988) Developmental regulation of a constitutively expressed mouse mRNA encoding a 72-kDa heat shock-like protein. Dev Biol 125:200–207

Giorda R, Ennis HL (1987) Structure of two developmentally regulated *Dictyostelium discoideum* ubiquitin genes. Mol Cell Biol 6:2097–2103

Glaser RL, Lis JT (1990) Multiple, compensatory regulatory elements specify spermatocyte-specific expression of the *Drosophila melanogaster* hsp26 gene. Mol Cell Biol 10:131–137

Glaser RL, Wolfner MF, Lis JT (1986) Spatial and temporal pattern of hsp26 expression during normal development. EMBO J 5:747–754

Graham RW, Jones D, Candido EPM (1989) UbiA, the major polyubiquitin locus in *Caenorhabditis elegans*, has unusual structural features and is constitutively expressed. Mol Cell Biol 9:268–277

Graziosi G, Micali F, Marzari R, De Cristini F, Savoini A (1980) Variability of response of early *Drosophila* embryos to heat shock. J Exp Zool 214:141–145

Greene JM, Larin Z, Taylor ICA, Prentice H, Gwinn KA, Kingston RE (1987) Multiple basal elements of a human hsp70 promoter function differently in human and rodent cell lines. Mol Cell Biol 7:3646–3655

Grinfeld S, Gilles J, Jacquet P, Baugnet-Mahieu L (1987) Late division kinetics in relation to modification of protein synthesis in mouse eggs blocked in the G2 phase after X-irradiation. Int J Radiat Biol 52:77–90

Haas I, Wabl M (1983) Immunoglobulin heavy chain binding protein. Nature (Lond) 306:387–389

Hahnel AC, Gifford DJ, Heikkila JJ, Schultz GA (1986) Expression of the major heat shock protein (hsp70) family during early mouse embryo development. Teratog Carcinog Mutagen 6:493–510

Haire RN, O'Leary JJ (1988) Mitogen-induced preferential synthesis of proteins during the Go to S phase transition in human lymphocytes. Exp Cell Res 179:65–78

Haire RN, Peterson MS, O'Leary JJ (1988) Mitogen activation induces the enhanced synthesis of two heat-shock proteins in human lymphocytes. J Cell Biol 106:883–891

Heikkila JJ, Schultz GA (1984) Different environmental stresses can activate the expression of a heat shock gene in rabbit blastocysts. Gamete Res 10:45–56

Heikkila JJ, Kloc M, Bury J, Schultz GA, Browder LW (1985a) Acquisition of the heat-shock response and thermotolerance during early development of *Xenopus laevis*. Dev Biol 107:483–489

Hensold JO, Housman DE (1988) Decreased expression of the stress protein HSP70 is an early event in murine erythroleukemic cell differentiation. Mol Cell Biol 8:2219–2223

Hensold JO, Hunt CR, Calderwood SK, Housman DE, Kingston RE (1990) DNA binding of heat shock factor to the heat shock element is insufficient for transcriptional activation in murine erythroleukemia cells. Mol Cell Biol 10:1600–1608

Hickey E, Brandon SE, Smale G, Lloyd D, Weber LA (1989) Sequence and regulation of a gene encoding a human 89-kilodalton heat shock protein. Mol Cell Biol 9:2615–2626

Hoffman EP, Corces VG (1984) Correct temperature induction and developmental regulation of a cloned heat shock gene transformed into the *Drosophila* germ line. Mol Cell Biol 4:2883–2889

Hoffman EP, Gerring SL, Corces VG (1987) The ovarian, ecdysterone, and heat-shock-responsive promoters of the *Drosophila melanogaster* hsp27 gene react very differently to perturbations of DNA sequence. Mol Cell Biol 7:973–981

Hoffmann EP, Corces V (1986) Sequences involved in temperature and ecdysterone-induced transcription are located in separate regions of a *Drosophila melanogaster* heat-shock gene. Mol Cell Biol 6:663673

Hoffmann T, Hovemann B (1988) Heat-shock proteins, Hsp84 and Hsp86, of mice and men: two related genes encode formerly identified tumour-specific transplantation antigens. Gene 74:491–501

Horrell A, Shuttleworth J, Colman A (1987) Transcript levels and translational control of hsp70 synthesis in *Xenopus* oocytes. Genes Dev 1:433–444

Horwich AL, Neupert W, Hartl F-U (1990) Protein-catalysed protein folding. TIBTECH 8:126–131

Howlett SK, Bolton VN (1985) Sequence and regulation of morphological and molecular events during the first cell cycle of mouse embryogenesis. J Embryol Exp Morphol 87:175–206

Howlett SK (1986) The effect of inhibiting DNA replication in the one-cell mouse embryo. Wilhelm Roux's Arch Dev Biol 195:499–505

Howlett SK, Barton SC, Surani MA (1987) Nuclear cytoplasmic interactions following nuclear transplantation in mouse embryos. Development 101:915–923

Hunter KW, Cook CL, Hayunga EG (1984) Leishmanial differentiation in vitro: induction of heat shock proteins. Biochem Biophys Res Comm 125:755–760

Iida H, Yahara I (1984) Durable synthesis of high molecular weight heat shock proteins in Go cells of the yeast and other eucaryotes. J Cell Biol 99:199–207

Ireland RC, Berger E, Sirotkin K, Yund MA, Osterbur D, Fristrom J (1982) Ecdysterone induces the transcription of four heat-shock genes in *Drosophila* S3 cells and imaginal discs. Dev Biol 93:498–507

Jansen-Durr P, Boeuf H, Kédinger C (1989) Cooperative binding of two E2F molecules to an Ela-responsive promoter is triggered by the adenovirus Ela, but not by a cellular Ela-like activity. EMBO J 8:3365–3370

Kaczmarek L, Calabretta B, Kao HT, Heintz N, Nevins J, Baserga R (1987) Control of hsp70 RNA levels in human lymphocytes. J Cell Biol 104:183–187

Kao HT, Capasso O, Heintz N, Nevins JR (1985) Cell cycle control of the human HSP70 gene: implications for the role of a cellular E1A-like function. Mol Cell Biol 5:628–633

Kao HT, Nevins JR (1983) Transcriptional activation and subsequent control of the human heat shock gene during adenovirus infection. Mol Cell Biol 3:2058–2065

Kelly SE, Cartwright IL (1989) Perturbation of chromatin architecture on ecdysterone induction of *Drosophila melanogaster* small heat shock protein genes. Mol Cell Biol 9:332–335

Kim KS, Kim YK, Lee AS (1990) Expression of the glucose-regulated proteins (GRP94 and GRP78) in differentiated and undifferentiated mouse embryonic cells and the use of the GRP78 promoter as an expression system in embryonic cells. Differentiation 42:153–159

Kim SH, He S, Kim JH (1984) Modification of thermosensitivity of HeLa cells by sodium butyrate, dibutyryl cyclic adenoside 3′:5′-monophosphate, and retinoic acid. Cancer Res 44:697–702

King ML, Davis R (1987) Do *Xenopus* oocytes have a heat shock response? Dev Biol 119:532–539

Kirschner M, Newport J, Gerhart J (1985) The timing of early developmental events in *Xenopus*. Trends Genet 1:41–47

Klemenz R, Gehring WJ (1986) Sequence requirement for expression of the *Drosophila melanogaster* heat shock protein hsp22 gene during heat shock and normal development. Mol Cell Biol 6:2011–2019

Kost SL, Smith DF, Sullivan WP, Welch WJ, Toft DO (1989) Binding of heat shock proteins to the avian progesterone receptor. Mol Cell Biol 9:3829–3838

Kothary R, Perry MD, Moran LA, Rossant J (1987) Cell-lineage-specific expression of the mouse hsp68 gene during embryogenesis. Dev Biol 121:342–348

Kothary R, Clapoff S, Darling S, Perry MD (1989) Inducible expression of an hsp68-lacZ hybrid gene in transgenic mice. Development 105:707–714

Krawczyk Z, Mąli P, Parvinen M (1988) Expression of a testis-specific hsp70 gene-related RNA in defined stages of rat seminiferous epithelium. J Cell Biol 107:1317–1323

Krone PH, Heikkila JJ (1988) Analysis of hsp30, hsp70 and ubiquitin gene expression in *Xenopus laevis* tadpoles. Development 103:59–67

Krone PH, Heikkila JJ (1989) Expression of microinjected hsp70/CAT and hsp30/CAT chimeric genes in developing *Xenopus laevis* embryos. Development 106:271–281

Kurtz S, Rossi J, Petko L, Lindquist S (1986) An ancient developmental induction in *Saccharomyces* sporulation and *Drosophila* oogenesis. Science 231:1154–1157

La Thangue NB, Rigby PWJ (1987) An adenovirus E1A-like transcription factor is regulated during the differentiation of murine embryonal carcinoma stem cells. Cell 49:507–513

La Thangue NB, Thimmapaya B, Rigby PWJ (1990) The embryonal carcinoma stem cell Ela-like activity involves a differentiation-regulated transcription factor. Nucl Acids Res 18:2929–2938

Lai BT, Chin NW, Stanek AE, Keh W, Lanks KW (1984) Quantitation and intracellular localization of the 85K heat shock protein by using monoclonal and polyclonal antibodies. Mol Cell Biol 4:2802–2810

Lambowitz AM, Kobayashi GS, Painter A, Medoff G (1983) Possible relationship of morphogenesis in pathogenic fungus, *Histoplasma capsulatum*, to heat shock response. Nature (Lond) 303:806–808

Landry J, Chrétien P, Lambert H, Hickey E, Weber LA (1989) Heat shock resistance conferred by expression of the human HSP27 gene in rodent cells. J Cell Biol 109:7–15

Lawrence F, Robert-Gero M (1985) Induction of heat shock and stress proteins in promastigotes of three *Leishmania* species. Proc Natl Acad Sci USA 82:4414–4417

Lee H, Simon JA, Lis JT (1988) Structure and expression of ubiquitin genes of *Drosophila melanogaster*. Mol Cell Biol 8:4727–4735

Lee SJ (1990) Expression of HSP86 in male germ cells. Mol Cell Biol 10:3239–3242

Legagneux V, Mezger V, Quélard C, Barnier JV, Bensaude O, Morange M (1989) High constitutive transcription of HSP86 gene in murine embryonal carcinoma cells. Differentiation 41:42–48

Levine RA, LaRosa GJ, Gudas LJ (1984) Isolation of cDNA clones for genes exhibiting reduced expression after differentiation of murine teratocarcinoma stem cells. Mol Cell Biol 4:2142–2150

Lindquist S (1986) The heat shock response. Annu Rev Biochem 55:1151–1191

Lindquist S, Craig EA (1988) The heat-shock proteins. Annu Rev Genet 22:631–677

Liu AYC, Bae-lee MS, Choi HS, Li B (1989a) Heat shock induction of HSP 89 is regulated in cellular aging. Biochem Biophys Res Comm 162:1302–1310

Liu AYC, Lin Z, Choi HS, Sorhage F, Li B (1989b) Attenuated induction of heat shock gene expression in aging diploid fibroblasts. J Biol Chem 264:12037–12045

Lowenhaupt K, Cartwright IL, Keene MA, Zimmerman JL, Elgin SCR (1983) Chromatin structure in pre- and postblastula embryos of *Drosophila*. Dev Biol 99:194–201

Maniak M, Nellen W (1988) A developmentally regulated membrane protein gene in *Dictyostelium discoideum* is also induced by heat shock and cold shock. Mol Cell Biol 8:153–159

Maresca B, Kobayashi GS (1989) Dimorphism in *Histoplasma capsulatum*: a model for the study of cell differentiation in pathogenic fungi. Microbiol Rev 53:186–209

Mason PJ, Hall LMC, Causz J (1984) The expression of heat-shock genes during normal development in *Drosophila melanogaster*. Mol Gen Genet 194:73–78

Matsumoto K, Uno I, Ishikawa T (1983) Initiation of meiosis in yeast mutants defective in adenylate cyclase and cyclic AMP-dependent protein kinase. Cell 32:417–423

Matsumoto M, Fujimoto H (1990) Cloning of hsp70-related gene expressed in mouse spermatids. Biochem Biophys Res Comm 166:43–49

Mestril R, Schiller P, Amin J, Klapper H, Ananthan J, Voellmy R (1986) Heat shock and ecdysterone activation of the *Drosophila melanogaster* hsp23 gene; a sequence element implied by developmental regulation. EMBO J 5:1667–1673

Mezger V, Bensaude O, Morange M (1987) Deficient activation of heat shock gene transcription in embryonal carcinoma cells. Dev Biol 124:544–550

Mezger V, Bensaude O, Morange M (1989) Unusual levels of HSE-binding activity in embryonal carcinoma cells. Mol Cell Biol 9:3888–3896

Mezquita J, Oliva R, Mezquita C (1987) New ubiquitin mRNA expressed during chicken spermiogenesis. Nucl Acids Res 15:9604

Milarski K, Welch WJ, Morimoto RI (1989) Cell cycle-dependent association of HSP70 with specific cellular proteins. J Cell Biol 108:413–423

Milarski KL, Morimoto RI (1986) Expression of human hsp70 during the synthetic phase of the cell cycle. Proc Natl Acad Sci USA 83:9517–9521

Morange M, Diu A, Bensaude O, Babinet C (1984) Altered expression of heat shock proteins in embryonal carcinoma and mouse early embryonic cells. Mol Cell Biol 4:730–735

Morgan WD (1989) Transcription factor Sp1 binds to and activates a human hsp70 gene promoter. Mol Cell Biol 9:4099–4104

Morgan WD, Williams GT, Morimoto RI, Greene J, Kingston RE, Tjian R (1987) Two transcriptional activators, CCAAT-binding transcription factor and heat shock transcription factor, interact with a human hsp70 gene promoter. Mol Cell Biol 7:1129–1138

Morganelli CM, Berger EM, Pelham HRB (1985) Transcription of *Drosophila* small hsp-tk hybrid genes is induced by heat shock and by ecdysterone in transfected *Drosophila* cells. Proc Natl Acad Sci USA 82:5865–5869

Morimoto R, Fodor E (1984) Cell-specific expression of heat shock proteins in chicken reticulocytes and lymphocytes. J Cell Biol 99:1316–1323

Morimoto RI, Tissieres A, Georgopoulos C (1990) Stress proteins in biology and medicine. Cold Spring Harbor Laboratory Press, New York

Muller WU, Li GC, Goldstein LS (1985) Heat does not induce synthesis of heat shock proteins or thermotolerance in the earliest stages of mouse embryo development. Int J Hyperthermia 1:97–102

Munro S, Pelham HRB (1986) An hsp70-like protein in the ER: identity with the 78 kd glucose-regulated protein and immunoglobulin heavy chain binding protein. Cell 46:291–300

Nakaki T, Deans RJ, Lee AS (1989) Enhanced transcription of the 78,000-Dalton glucose regulated protein (GRP78) gene and association of GRP78 with immunoglobulin light chain in a non-secreting B-cell myeloma line (NS-1). Mol Cell Biol 9:2233–2238

Neidhardt FC, VanBogelen RA (1987) Heat shock response. In: Neidhardt FC (ed) *Escherichia coli* and *Salmonella typhimurium*. Cellular and molecular biology, vol 2. American Society for Microbiology, Washington, DC, pp 1334–1345

Nene V, Dunne DW, Johnson KW, Taylor DW, Cordingley JS (1986) Sequence and expression of a major egg antigen from *Schistosoma mansoni*. Homologies to heat shock proteins and alpha-crystallins. J Biol Chem 21:179–188

Neupert W, Hartl FU, Craig EA, Pfannert N (1990) How do polypeptides cross the mitochondrial membranes? Cell 63:447–450

Nevins JR (1982) Induction of the synthesis of a mammalian 70 kd heat shock protein by the adenovirus Ela gene product. Cell 29:913–919

Nickells RW, Browder LW (1985) Region-specific heat-shock protein synthesis correlates with a biphasic acquisition of thermotolerance in *Xenopus laevis* embryos. Dev Biol 112:391–395

Nickells RW, Browder LW (1988) A role for glyceraldehyde-3-phosphate dehydrogenase in the development of thermotolerance in *Xenopus laevis* embryos. J Cell Biol 107:1901–1909

Onclercq R, Gilardi P, Lavenu A, Cremisi C (1988) c-*myc* products *trans*-activate the adenovirus E4 promoter in EC stem cells by using the same target sequence as E1A products. J Virol 62:4533–4537

Ovsenek N, Heikkila JJ (1990) DNA sequence-specific binding activity of the heat-shock transcription factor is heat-inducible before the midblastula transition of early *Xenopus* development. Development 110:427–433

Palter KB, Watanabe M, Stinson L, Mahowald AP, Craig EA (1986) Expression and localization of *Drosophila melanogaster* hsp70 cognate proteins. Mol Cell Biol 6:1187–1203

Parsell DA, Sauer RT (1989) Induction of a heat shock-like response by unfolded protein in *Escherichia coli*: dependence on protein level not protein degradation. Genes Dev 3:1226–1232

Pelham HRB (1986) Speculations on the functions of the major heat shock and glucose-regulated proteins. Cell 46:959–961

Pelham HRB (1989) Heat shock and sorting of luminal ER proteins. EMBO J 8:3171–3176

Petko L, Linquist S (1986) Hsp26 is not required for growth at high temperature, nor for thermotolerance, spore development, or germination. Cell 45:885–894

Poueymirou WT, Schultz RM (1989) Regulation of mouse preimplantation development: inhibition of synthesis of proteins in the two-cell embryo that require transcription by inhibitors of cAMP-dependent protein kinase. Dev Biol 133:588–599

Raychaudhuri P, Rooney R, Nevins JR (1987) Identification of an E1A-inducible cellular factor that interacts with regulatory sequences within the adenovirus E4 promoter. EMBO J 6:4073–4081

Rebbe NF, Hickman WS, Ley TJ, Stafford DW, Hickman S (1989) Nucleotide sequence and regulation of a human 90-kDa heat shock protein gene. J Biol Chem 264:15006–15011

Reichel R, Kovesdi I, Nevins JR (1987) Developmental control of a promoter-specific factor that is also regulated by the E1A gene product. Cell 48:501–506

Richards FM, Watson A, Hickman JA (1988) Investigation of the effects of heat shock and agents which induce a heat shock response on the induction of differentiation of HL-60 cells. Cancer Res 48:6715–6720

Riddihough G, Pelham HRB (1986) Activation of the *Drosophila* hsp27 promoter by heat shock and by ecdysterone involves independent and remote regulatory sequences. EMBO J 5:1653–1658

Rimland J, Akhayat O, Infante D, Infante AA (1988) Developmental regulation and biochemical analysis of a 21 Kd heat shock in sea urchins. J Cell Biochem Suppl 12D:271

Roccheri MC, Di Bernardo MG, Giudice G (1981) Synthesis of heat-shock proteins in developing sea urchins. Dev Biol 83:173–177

Rosen E, Sivertsen A, Firtel RA (1983) An unusual transposon encoding heat shock inducible and developmentally regulated transcripts in dictyostelium. Cell 35:243–251

Rothman JE (1989) Polypeptide chain binding proteins: catalysts of protein folding and related processes in cells. Cell 59:591–601

Ruder FJ, Frasch M, Mettenleiter TC, Büsen W (1987) Appearance of two maternally directed histone H2A variants precedes zygotic ubiquitination of H2A in early embryogenesis of *Sciara coprophila* (Diptera). Dev Biol 122:568–576

Santoro MG, Garaci E, Amici C (1989) Prostaglandins with antiproliferative activity induce the synthesis of a heat shock protein in human cells. Proc Natl Acad Sci USA 86:8407–8411

Savoini A, Micali F, Marzari R, De Cristini F, Graziosi G (1981) Low variability of the protein species synthesized by *Drosophila melanogaster* embryos. Wilhelm Roux's Arch Dev Biol 190:161–167

Shapira M, McEwen JG, Jaffe CL (1988) Temperature effects on molecular processes which lead to stage differentiation in *Leishmania*. EMBO J 7:2895–2901

Shin DY, Matsumoto K, Iida H, Uno I, Ishikawa T (1987) Heat shock response of *Saccharomyces cerevisiae* mutants altered in cyclic AMP-dependent protein phosphorylation. Mol Cell Biol 7:244–250

Shyamala G, Gauthier Y, Moore SK, Catelli MG, Ullrich SJ (1989) Estrogenic regulation of murine uterine 90-kilodalton heat shock protein gene expression. Mol Cell Biol 9:3567–3570

Simon MC, Fisch TM, Benecke BJ, Nevins JR, Heintz N (1988) Definition of multiple, functionally distinct TATA elements, one of which is a target in the hsp70 promoter for E1A regulation. Cell 52:723–729

Simon MC, Kitchener K, Kao HT, Hickey E, Weber L, Voellmy R, Heintz N, Nevins JR (1987) Selective induction of human heat shock transcription by the adenovirus E1A gene products, including the 12S E1A product. Mol Cell Biol 7:2884–2890

Singh MK, Yu J (1984) Accumulation of a heat shock-like protein during differentiation of human erythroid cell line K562. Nature (Lond) 309:631–633

Sirotkin K, Davidson N (1982) Developmentally regulated transcription from *Drosophila melanogaster* chromosomal site 67B. Dev Biol 89:196–210

Suemori H, Hashimoto S, Nakatsuji N (1988) Presence of the adenovirus E1A-like activity in preimplantation stage mouse embryos. Mol Cell Biol 8:3553–3555

Susek RE, Lindquist SL (1989) hsp26 of *Saccharomyces cerevisiae* is related to the superfamily of small heat shock proteins but is without a demonstrable function. Mol Cell Biol 9:5265–5271

Taylor ICA, Kingston Re (1990) Factor substitution in a human hsp70 gene promoter: TATA-dependent and TATA-independent interactions. Mol Cell Biol 10:165–175

Taylor KD, Piko L (1987) Patterns of mRNA prevalence and expression of B1 and B2 transcripts in early mouse embryos. Development 101:877–892

Theodorakis NG, Zand DJ, Kotzbauer PT, Williams GT, Morimoto RI (1989) Hemin-induced transcriptional activation of the HSP70 gene during erythroid maturation in k562 cells is due to a heat shock factor-mediated stress response. Mol Cell Biol 9:3166–3173

Ting LP, Tu CL, Chou CK (1989) Insulin-induced expression of human heat-shock protein gene hsp70. J Biol Chem 264:3404–3408

Ullrich SJ, Robinson EA, Law LW, Willingham M, Appella E (1986) A mouse tumor-specific transplantation antigen is a heat shock-related protein. Proc Natl Acad Sci USA 83:3121–3125

Ullrich SJ, Moore SK, Appella E (1989) Transcriptional and translational analysis of the murine 84- and 86-kDa heat shock proteins. J Biol Chem 264:6810–6816

Van der Ploeg LHT, Giannini SH, Cantor CR (1985) Heat shock genes: regulatory role for differentiation in parasitic protozoa. Science 228:1443–1446

Vasseur M, Condamine H, Duprey P (1985) RNAs containing B2 repeated sequences are transcribed in the early stages of mouse embryogenesis. EMBO J 7:1749–1753

Vitek MP, Berger EM (1984) Steroid and high-temperature induction of the small heat-shock protein gene in *Drosophila*. J Mol Biol 178:173–189

Voellmy R, Rungger D (1982) Transcription of a *Drosophila* heat shock gene is heat-induced in *Xenopus* oocytes. Proc Natl Acad Sci USA 79:1776–1780

Werner-Washburne M, Becker J, Kosic-Smithers J, Craig EA (1989) Yeast Hsp70 RNA Levels vary in response to the physiological status of the cell. J Bact 171:2680–2688

White RJ, Stott D, Rigby PWJ (1989) Regulation of RNA polymerase III transcription in response to F9 embryonal carcinoma stem cell differentiation. Cell 59:1081–1092

Williams GT, McClanahan TK, Morimoto RI (1989) Ela transactivation of the human HSP70 promoter is mediated through the basal transcriptional complex. Mol Cell Biol 9:2574–2587

Wittig S, Hensse S, Keitel C, Elsner C, Wittig B (1983) Heat shock gene expression is regulated during teratocarcinoma cell differentiation and early embryonic development. Dev Biol 96:507–514

Wu B, Hunt C, Morimoto R (1985a) Structure and expression of the human gene encoding major heat shock protein HSP70. Mol Cell Biol 5:330–341

Wu BJ, Morimoto RI (1985b) Transcription of the human hsp70 gene is induced by serum stimulation. Proc Natl Acad Sci USA 82:6070–6074

Wu BJ, Hurst HC, Jones NC, Morimoto RI (1986a) The E1A 13S product of adenovirus-5 activates transcription of the cellular human hsp70 gene. Mol Cell Biol 6:2994–2999

Wu BJ, Kingston RE, Morimoto RI (1986b) Human hsp70 promoter contains at least two distinct regulatory domains. Proc Natl Acad Sci USA 83:629–633

Wu BJ, Williams GT, Morimoto RI (1987) Detection of three protein binding sites in the serum regulated promoter of the human gene encoding the 70 kDa heat shock protein. Proc Natl Acad Sci USA 84:2203–2207

Xiao H, Lis JT (1989) Heat shock and developmental regulation of the *Drosophila melanogaster* hsp83 gene. Mol Cell Biol 9:1746$_{1753}$

Young RA (1990) Stress proteins and immunology. Annu Rev Immunol 8:401–420

Yufu Y, Nishimura J, Ideguchi H, Nawata H (1990) Enhanced synthesis of heat-shock proteins and augmented thermotolerance after induction of differentiation in HL-60 human leukemia cells. FEBS Lett 268:173–176

Yufu Y, Nishimura J, Takahira H, Ideguchi H, Nawata H (1989) Down-regulation of a M_r 90,000 heat shock cognate protein during granulocytic differentiation in HL-60 human leukemia cells. Cancer Res 49:2405–2408

Zakeri ZF, Wolgemuth DJ (1987) Developmental-stage-specific expression of the hsp70 gene family during differentiation of the mammalian germ line. Mol Cell Biol 7:1791–1796

Zakeri ZF, Ponzetto C, Wolgemuth DJ (1988a) Translational regulation of the novel haploid-specific transcripts for the c-abl proto-oncogene and a member of the 70 kDa heat-shock protein gene family in the male germ line. Dev Biol 125:417–422

Zakeri ZF, Wolgemuth DJ, Hunt CR (1988b) Identification and sequence analysis of a new member of the mouse HSP70 gene family and characterization of its unique cellular and developmental pattern of expression in the male germ line. Mol Cell Biol 8:2925–2932

Zimarino V, Tsai C, Wu C (1990) Complex modes of heat shock factor activation. Mol Cell Biol 10:752–759

Zimmerman JL, Petri W, Meselson M (1983) Accumulation of a specific subset of *D. melanogaster* heat-shock mRNAs in normal development without heat shock. Cell 32:1161–1170

Zuker C, Cappello J, Lodish HF, George P, Chung S (1984) *Dictyostelium* transposable element DIRS-1 has 350-base-pair inverted terminal repeats that contain a heat shock promoter. Proc Natl Acad Sci USA 81:2660–2664

The Interactions of Water and Proteins in Cellular Function

J.G. WATTERSON[1]

1 Introduction

Our understanding of the role of water in biology is extremely limited; this most abundant component of living cells is traditionally viewed as structureless, space-filling, background medium in which biochemical events occur. The biochemical and biophysical reactions occurring in aqueous regions of the cell are viewed as occurring in aqueous solution. However, this view disregards the predominantly gelled state of the cell interior. Kempner and Miller (1968) showed by centrifuging intact cells that the fluid aqueous portion of the cytoplasm is devoid of macromolecules. Every biochemist knows that spinning a protein solution under native conditions at 300 000 g, corresponding to many hundreds of atmospheres pressure, does not yield a protein concentration as high as 5%. These observations demonstrate that intracellular proteins in their native conformations are able to gel a large proportion of cellular water and prevent its flow.

That extracted enzymes function in solution in vitro is a fact that has traditionally been extrapolated into the concept of the living cell by most biochemists. However, there is no justification for this view, indeed there is a great deal of evidence that speaks against it, as thoroughly reviewed by Welch (1977). In the alternative view, enzymes are associated into complex protein networks permeating the space of the cytoplasm. The existence of such superstructures means that the sequence of chemical steps in metabolic pathways has a corresponding physical state in the cell, wherein enzymes are organized in that sequence in large subcellular aggregates (Srere 1987; Srivastrava and Bernhard 1987). It further means that old concepts of enzyme kinetics and random diffusion of substrate molecules are not appropriate in the reality of the cell (Masters et al. 1987) and implies flexible but highly regulated association-dissociation equilibria and translocation of enzymes themselves (Kaprelyants 1988). This picture of extensive three-dimensional order is strongly supported by the cytoplasmic matrix, as revealed by the work of Porter and his group (1983). But it also poses a problem: such an intricate network of protein scaffolding would hinder, rather than help, intracellular streaming and the free diffusion of metabolites. In other words, the concepts of efficient cytoplasmic movement and structure appear to be mutually exclusive. This question is at present under investi-

[1]Department of Science, Gold Coast University College, Private Mail Bag 50, Gold Coast, Queensland 4217, Australia

gation (Gershon et al. 1985; Luby-Phelps et al. 1988). As argued by Clegg (1984), all the components of the cell, including water, "should be considered as a single system if we are to understand the whole." Thus, the matrix and aqueous elements must work together, as opposed to independently, to produce cytoplasmic movements.

It is now well established that the cytoskeleton is composed of long, linear filaments involved in the production of mechanical force and direction of movement within the cell. Thus, contraction (Huxley 1973), cytoplasmic gel-sol transition and streaming (Taylor and Fechheimer 1987) and axonal transport (Amos and Amos 1985) occur via energetic protein interactions with these filaments. These models are today widely accepted, but in none of them is any significant role assigned to water. If we fill the subcellular space with a simple, liquid solvent, we introduce difficulties for these models from the point of view of the traditional concept of the essential randomness of molecular events. For these processes to function properly, the enzymic machinery that converts chemical energy into work is required to cycle through a series of precise physical steps which cannot tolerate disruptive energetic bombardment from outside. The energy source, usually a phosphate bond of ATP, is equivalent to 10 or 20, or at the very maximum 30, hydrogen bonds and so could hardly be used to tame the violent surroundings of many hundreds of independently acting water molecules, let alone also then be used for the task at hand. A resolution of this problem is offered by a concept of co-operation and participation on the part of the solvent in the function of molecular machinery (Wiggins and MacClement 1987). Examples of physical aspects of such a new concept are the structured interfacial or vicinal water (Drost-Hansen 1985) and the liquid crystalline nature of protein gels (Buxbaum et al. 1987). Recent nuclear magnetic resonance (Lamanna and Cannistraro 1989) and neutron scattering (Giordano et al. 1990) studies on dilute protein solutions have revealed long-range solute-solvent interactions, supporting the idea of solvent involvement.

According to the principles of statistical thermodynamics, the probability that a polymer the size of a protein will spontaneously fold into a unique conformation is vanishingly small if the overall "stabilizing" energy is not very large. But this stabilizing energy is not at all large, being about 10 kcal/mol (Kauzmann 1959; Tanford 1968 and Privalov 1979), which is near that of a phosphate or 10 to 20 hydrogen bonds. In addition, the substitution of a single amino acid can reduce this value by as much as 3 kcal/mol (Goldenberg 1988). Such information has often led to the conclusion that proteins are not very stable, thermodynamically speaking, and raised the question whether they can really adopt a single conformation, when thermodynamics dictates the "kicking and screaming stochastic molecule" of the statistical world (Weber 1975). This problem was addressed by Kauzmann (1959), who proposed the "hydrophobic bond" as the mechanism that gives globular proteins their stability; this bond operating in the protein interior between nonpolar side groups which are repelled by water but attracted to one another. The idea was vigorously criticized, for example by Hildebrand (1968), and during the 1970's the concept was renamed with the terms "hydrophobic effect" and "hydrophobic interaction" (Tanford 1973; Franks 1975). These latter terms imply only a thermodynamic preference shown by nonpolar residues for a hydrocarbon over an aqueous surrounding and do not portray a mechanical picture of protein stability.

The number of energy states readily available to a protein molecule is extremely large, 10^x where x lies between 30 and 70 (Frauenfelder (1983) suggests that x is around 50). Since a protein does not possess special conformational states which differ greatly in energy from one another, the question arises as to whether it exists in all these 10^x states. There is not yet agreement on the answer, as can be seen from directly contradictory opinions expressed in recent reviews (Frauenfelder et al. 1988; Goldenberg 1988). One strategy often used assumes that proteins adopt only those states which are virtually equal in energy to that of the native active state. However, the two principle conformations adopted by hemoglobin, tense and relaxed, are separated by 3 kcal/mol (Perutz 1979), a very sizable fraction of its overall stabilizing energy; so one cannot regard these conformations as energetically close, for otherwise so is the unfolded conformation! On the other hand, one might argue that the energy barriers (activation energies) to most of the 10^x states are too high to be surmounted and consequently only a few states can be occupied. This is a common view of workers interested in protein folding, since this process appears to occur via a strict kinetic sequence of a few conformational intermediates. But such barriers must truly be extraordinarily high, amounting to hundreds of kcal/mol, because the average fluctuation in energy due to random thermal motion in a representative molecule of 25 000 Da exceeds 30 kcal/mol (Cooper 1976). We would then expect that all proteins must have intramolecular bonding arrangements of a very specialized nature indeed to ensure the existence of such high barriers.

The subject of random thermal fluctuations in proteins has been under investigation for more than 20 years. Welch et al. (1982) published an excellent overview of prominent models based on the classical approach, to which must be added the newer computer simulations which are gaining much attention (Karplus and McCammon 1983). The number of three-dimensional protein structures deduced from X-ray crystallography exceeds 200 and the prediction of new ones from sequence data is beginning to show success (Crawford et al. 1987), yet there is no evidence so far of structural channelling mechanisms which could collect and divert the random energy to do useful work at the active site. A different approach proposes that this energy can be stored in rarely occurring, irreversible transitions (Kell 1988).

The idea that molecular fluctuations can perform work is reminiscent of the popular theory of osmosis, in which the random motion of molecules is proposed to provide a force. This mechanism is illustrated in Fig. 1, in which a semi-permeable membrane separates solvent and solution phases. There are fewer solvent molecules (circles) in the solution than in the pure solvent because some have been displaced by solute molecules (stars). Solvent molecules move randomly in all directions, and initially the pressure on both sides of the membrane is equal. Fewer molecules pass from the solution into the solvent than in the opposite direction because there are fewer of them to collide with the membrane of the solution side. This results in a net flow of solvent into the solution, increasing the pressure there above that on the solvent side. The flow against ever-increasing pressure continues until the osmotic

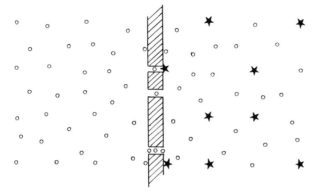

Fig. 1. Pictorial scheme showing a semi-permeable membrane separating solvent on the *left* from solution on the *right*, whereby the solvent molecules (*circles*) can move through the pores but the solutes (*stars*) cannot. Of course the circles should be everywhere in total contact, filling the whole space and not separated as shown here. According to the "molecular theory" of osmosis, the solvent molecules on the *right* are fewer in number, but are diffusing faster than their counterparts on the *left* because of the higher pressure in the solution

pressure is reached, i.e., the pressure difference across the membrane which stops this solvent net flow. The higher pressure in the solution increases the rate at which solvent molecules collide with the membrane, compensating for their fewer number, so the total number of solvent molecules passing from each side is now equal.

Not all physicists and chemists have supported the above theory. In their well-known text, Glasstone and Lewis (1963) give an extensive account of at least five mechanisms of osmosis proposed since Van't Hoff published his law a century ago, and it is of interest to note that none of them is deemed satisfactory by those authors. More recently, other "uphill" flow phenomena have also been detected (Gaeta and Mita 1979). As outlined above, the commonly accepted theory presents solvent flowing against pressure as a result of random molecular motion. However, the displacement of this mass by such a process would break Newton's Second Law of Motion, because the force acting on it during its acceleration is in the wrong direction. One must therefore conclude that the force underlying osmosis does not push the solvent into the solution, but must pull it. This conclusion demands a radical shift in our view of molecular events away from the old concepts of randomness.

2 The Cluster Model of Liquid Structure

2.1 Cluster Size

The idea that tension exists in liquids is not a new one. Its historical development can be found in the excellent review of Hammel and Scholander (1976). In fact, these authors themselves have proposed (Scholander et al. 1965) a modern version of the theory, in which osmosis is caused by the enhancement of "solvent tension" by solute. However, their theory has been widely criticized (Plumb and Bridgman 1972; Andrews 1976; Hildebrand 1979). Their model is difficult to visualize in molecular terms, because, just as with the statistical theories of liquid structure, it relies also on random collisions. It is proposed that these interactions can give rise to tension. Thus, the basis of their mechanism appears to involve a contradiction, since collisions imply pressure and not tension.

This difficulty does not arise in the cluster model, which is based on the concept of dynamic, co-operative grouping of liquid molecules resulting from intermolecular bonding (Watterson 1982). In this picture tension can be exerted in any one direction as far as the molecules are interconnected in that direction. It is self-evident that tension cannot extend beyond a break in these connections and so, over spatial dimensions larger than those within which unbroken interconnections extend, pressure, and not tension, operates. In other words, at a given instant, tension is felt over a region of space that is as large as a cluster. This picture does not require that every possible bond is formed, as exemplified by regions with ice structure forming within liquid water. It means only that an unbroken interconnection, percolating through the cluster from one side to the other, exists at any instant.

An important aspect of the model is that the making and breaking of bonds are co-operative processes. This rules out the idea that clusters have flickering existence, appearing and disintegrating spontaneously at random. On the contrary, they move about because the change in the bonded state of a group of molecules affects that of its neighbors, and thus the making and breaking processes travel as on-going polymerization and depolymerization reactions through the liquid medium. These processes do not stop and restart independently at random, but are continuous and add together to give a wave motion. We have now the picture of a structure wave completely filling the liquid space, so there is no region where cluster formation is not occurring. The dimensions of a cluster are those than define the wave motion, i.e., the wavelengths. In the case of pure bulk liquid without boundaries this medium is isotropic and so can be idealized as a three-dimensional array of cubic wave units, each defining a cluster of volume u_0. At the corners of the units where the clusters meet are the point nodes in the wave motion. When a foreign solute molecule is introduced into the solvent it disrupts the solvent-solvent interactions which underlie the motion of the structure wave, whatever the particular nature of the solute-solvent interactions. As a result, an extra node forms at the position of the solute, just as an obstacle in any vibrating system generally produces a node. These extra nodes shorten the wavelength, and as a consequence decrease the volume of the wave units to u.

We can now extend this model to explain some fundamental properties of solutions by assuming that it is these wave units or clusters, rather than the single solvent molecules, that act as individual entities. In other words, a cluster can be viewed in some respects like a particle, because the molecules that comprise it at any given moment are bonded together. This picture leads directly to a simple explanation of the colligative properties of solutions (Watterson 1987a,b). For example, the assumption that the clusters evaporate as entire units gives a simple formula for the reduction of vapor pressure of a solution compared to the pure solvent. Since clusters are smaller in the solution, fewer solvent molecules enter the vapor and so the pressure is correspondingly reduced by the ratio

$$\frac{P}{P_0} = \frac{u}{u_0},\tag{1}$$

where P and P_0 are the vapor pressures of the solution and pure solvent respectively.

Figure 2 shows how different size clusters interact to produce osmotic equilibrium. In the solution, where the cluster size is reduced, shorter wavelength corresponds to higher tension, just as higher pitch of a vibrating string corresponds to higher tension. This difference has an effect in the region where clusters of different sizes are in contact, in that the material in the smaller clusters exerts a net pull on that in the larger. This is a force operating from within these structures. It is a mechanical force, and if equilibrium is not yet reached, there is transfer of solvent into the solution. In other words, the force results from collective molecular action, and not from independent random collisions.

Since the solvent medium pervades both phases, the wave is propagated in both directions across the membrane. In other words, clusters must also move back and forth across the membrane. They should not be viewed as static structures, as may be falsely interpreted from Fig. 2. Although they change size as they cross this boundary, they must not change energy when the system reaches equilibrium. Then there can be exchange, unit for unit, across the boundary without any further transfer of energy. This argument leads to the conclusion that the clusters obey the Gas Law

$$P_0 u_0 = kT,\tag{2}$$

where k is Boltzmann's constant or the gas constant expressed per molecule (Watterson 1987a). This means that at room temperature and pressure u_0 is about 40 nm^3,

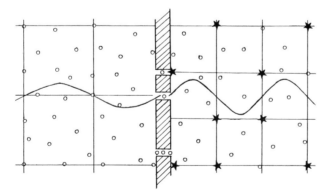

Fig. 2. Schematic representation of Fig. 1, now showing solvent clusters defined by the structure wave. In the solution, the solute molecules (*stars*) are located at nodes on the grid of the stationary wave pattern, and the higher tension in the smaller cluster is shown by the larger amplitude of the shorter wave. At equilibrium, the wave, i.e., clusters, passes smoothly in both directions accross the boundary unit for unit, changing its wavelength and amplitude but not its energy

i.e., a cube with an edge about 3.4 nm long. In the case of water this distance is spanned by 11 molecules roughly, so that a cluster contains 1300 to 1400 molecules and has a molecular mass of about 24 000 Da.

2.2 Cluster Energetics

We are all familiar with the ability of osmotic systems to do work. In terms of the cluster model, this happens while the two phases in contact, pure solvent and solution, move spontaneously towards equilibrium. The larger, energy-rich clusters carry energy into the solution phase until the smaller clusters acquire an equal amount of energy as given by Eq.(2). Clusters can exchange their energies and in doing so provide work, and this idea is expanded here to establish the quantitative relationship between cluster energy and work.

The upper panel in Fig. 3 represents the cycling of a piston in a cylinder just as in the classical Carnot cycle (Moore 1956). The wall of this cylinder contains an opening with a shutter, positioned opposite the piston, which allows the entry and

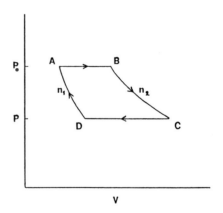

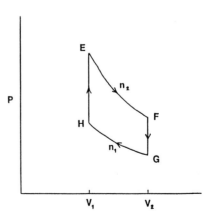

Fig. 3. Work cycles driven by a colligative potential. Heat alone is absorbed by the machine when it contains n_2 molecules of vapor and returned to the environment when it contains n_1 molecules, the difference being converted into work

removal of gases. Two large systems are available to act as source and sink. The first, the source, is a closed volume containing a pure liquid in equilibrium with its vapor at pressure P_0 for temperature T. The second, the sink, contains a dilute ideal solution of a nonvolatile solute in the liquid with the vapor now at the reduced pressure P for the same temperature. In addition, the environment, i.e., any large body also at this temperature, can act as source or sink for the transfer of heat alone. At A, the machine, already containing n_1 molecules of vapor, is placed in contact with the vapor above the pure solvent with the cylinder open. The piston does work reversibly by moving against pressure P_0 while (n_2-n_1) molecules of gas enter from the vapor source. Then at B, the cylinder is closed and the piston now expands down the isotherm, i.e., the familiar PV hyperbolic curve, while absorbing the necessary heat from the environment. From C to D and D to A the steps are reversed with the machine open to the vapor at the lower pressure, P, above the solution (C to D) and then closed (D to A).

The work terms given by the areas under the isobars must be equal since the (n_2-n_1) molecules of gas that entered from the source are all returned to the sink at the same temperature and thus take out with them the energy they brought into the machine. The work done by the machine is, therefore, given by the difference in areas under the isotherms

$$W = (n_2 - n_1) \, kT \, \ln (P_0/P). \tag{3}$$

The lower panel in Fig. 3 shows the symmetrical cycle obtained by interchanging the P,V axes. In this case the contents of the cylinder are diluted by a step between two set volumes instead of two set pressures. At the start, E, the machine already contains n_2 molecules of gas and, with the shutter closed, it expands down the isotherm to F while absorbing heat. It is then placed in contact with the vapor of a series of sources in succession and the shutter is opened each time to allow equilibration of the gas in the machine with each source in turn without further increase in volume. The vapor pressure of each source is maintained at a slightly lower value than that of the preceding source by having a corresponding incremental increase in solute concentration. At G there are n_1 molecules left in the machine, and from G to H and H to E the steps are reversed to complete the cycle. Here again, the work done is given by the area under the isotherms.

$$W = (n_2 - n_1) \, kT \, \ln (V_2/V_1). \tag{4}$$

These elementary thermodynamic cycles yield the result expected from classical thermodynamics: namely, that the work performed by the machine equals the free energy of dilution, whether the dilution step is an expansion between two given pressures or two given volumes. But, most importantly, the energy used to do this work is the difference in the heat exchanged by the machine while the shutter is closed, and does not originate from the solutions.

The machine can, however, operate in another cycle, shown in Fig. 4, using the steps at constant pressure and volume only, thereby eliminating any heat exchanges. At A the cylinder, already containing vapor at pressure P_0, is placed in contact with vapor above the pure solvent. The shutter is opened and the piston expands revers-

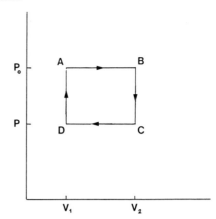

Fig. 4. Rectangular work cycle of the machine operating between the vapor pressure of the pure solvent P_0 and its colligatively reduced pressure P

ibly to B while the required number of molecules, $P_0(V_2-V_1)/kT$, enter the machine. From B to C, the contents are equilibrated with the sources of gradually increasing solute concentration in succession until the pressure falls to P. From C to D and D to A the steps are reversed to complete the cycle performing the work

$$W = (P_0 - P)(V_2 - V_1). \tag{5}$$

The work done per molecule is thus

$$w = \frac{P_0 - P}{P_0} kT, \tag{6}$$

where $(P_0-P)/P_0$ is that fraction of molecules which leaves the machine during the drop in pressure, B to C, and so are not expelled by the return stroke, C to D, when work is done on the machine. All the molecules that enter during the expansion at P_0 are transferred to sources at lower pressures, but because they enter and leave at the same temperature, they take out with them all the energy they brought into the machine. However, in contrast to the cycles in Fig. 3, where work is obtained from heat, in this case the energy must be supplied by a nonthermal source within the solutions. This conclusion is a result of thermodynamic argument and is therefore quite independent of the cluster model.

To my knowledge there is no published interpretation of the rectangular cycle, either in terms of classical thermodynamics or presentday statistical theories of liquids. In terms of the cluster model, however, the reason for the availability of energy is clear. According to Eq.(1), the vapor pressures of the sources are directly proportional to the cluster sizes in their corresponding solutions, and so Eq.(6) becomes

$$w = \frac{u_0 - u}{u_0} kT. \tag{7}$$

Thus, the energy converted into work is equivalent to the drop in size of clusters between the first and the last vapor source. This result reinforces the notion of the cluster as a wave unit, i.e., the unit of a vibrating system, because in such a picture clusters have something of the character of springs. In the pure solvent the spring is strongest, while in the solution it is weaker or less energetic, unit for unit.

In discussing cluster size, bulk liquid was depicted as being composed of a three-dimensional array of cubic clusters produced by an infinite isotropic stationary form of the structure wave. This picture is, of course, idealized, because in the absence of boundaries there are no given directions to guide wave propagation and reflection, and so the motion would not become stationary and form a pattern. One of the simplest examples of a boundary is an infinite two-dimensional surface, such as the air-liquid or solid-liquid interface. This surface would force a stationary two-dimensional planar node to take up position in the wave motion. The formation of such a node need not depend on the chemical nature of the solute surface, because its very existence would induce those molecules in the monolayer adjacent to it to have altered motions compared to those in the bulk. A nodal plane defines a face or side of a cluster, and so the presence of the surface would induce the side-by-side alignment of liquid clusters to spread over it. In this arrangement they form a layer of solvent, one cluster thick, adjacent to the interface.

Over the past two decades, surface hydration effects have been the subject of ever-increasing research interest. Their underlying cause is disputed, but the fact they are observed at clay, organic, and biological surfaces means that they must be a property of water, rather than the solute or the interface. An important effect is the presence of a repulsive "hydration force" exerted outward, normal to the solute surface, the molecular origin of which is an extremely controversial topic (Churaev and Derjaguin 1985). It has been measured in various experimental systems including clays (Norrish 1954; Van Olphen 1954), lipid bilayers (LeNeveu et al. (1976) and mica sheets (Israelachvilli and Adams 1976), and considering the diversity in the chemical nature of the surface investigated, it is astounding how often the distance of 3 to 4 nm is found to be their range of influence (Ninham 1980). The existence of a strong, lateral, inward tension, operating at right angles to the applied force, explains in a simple way the power exhibited by solute systems stacked in layers, such as hydrophilic clays or lamellar micelles, when they swell against imposed pressure (Watterson 1989).

Figure 5 illustrates a simplified view of cluster dynamics. It shows how boundaries may induce harmonic transitions to produce clusters of any size. It also shows how a stationary boundary forming a stationary wave results in a stationary cluster. This does not mean stationary molecules; they move just as in bulk water and maintain the wave motion. Only the monolayer of water molecules adsorbed directly onto the solute surface need be restricted in their movement.

On the other hand, stationary clusters mean no bulk flow. Thus in regions where stationary clusters are aligned together the medium forms a gel. There is no macroscopic flow of solvent because the clusters are fixed in space by being anchored onto the surfaces of fixed solutes. In this picture the packing of solutes, i.e., protein and lipid assemblies, together with stationary clusters is responsible for the gelled state of the cell interior. This is a different view from that offered by the popular theory of gelation. This latter theory, refined by Flory (1953), is based on the infinite degree of cross-linking of a polymeric solute present in the medium. Al-

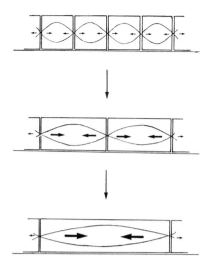

Fig. 5. Clusters become aligned in a row on a flat surface because the presence of the surface guides the direction of the wave motion parallel and normal to it, thus forming a fixed planar node in the wave pattern. Harmonic transitions can then convert many small clusters into a single large one. At the same time the internal lateral tension shown by the *arrows* is exerted over longer distances

though the theory explains why macromolecules stop flowing relative to one another, it has no role for the solvent, which in some systems can be as high as 99% of total content. But it is precisely the cessation of flow of solvent, and not solute, that is so surprising when a gel sets. Thus, to understand the phenomenon so pertinent to the subcellular world, we need a picture in which solvent interactions are seen as playing the central role. This model must explain how small solvent molecules with their individual liquid motions can become interconnected over distances that are long enough to prevent macroscopic flow. Since the popular theory is based solely on the statistical properties of polymeric solutes, it cannot be considered as an explanation of gelation in molecular terms.

3 The Domain Model of Protein Structure

3.1 Domain Size

Why are proteins the size they are? This question has become particularly intriguing since it has been recognized that globular proteins are folded into discrete domains (Richardson 1981; Rossmann and Argos 1981). Table 1 lists 16 proteins composed of a single chain folded into a single domain whose three-dimensional structures are known from X-ray crystallography. Each entry is a member of a large class of similar proteins in terms of tertiary structure. The average length of the 16 listed examples is 192 residues, and most fall within the range 150 to 250.

The citations are listed in chronological order so that there is no attempt at classification. However, prominence should be given to triose phosphate isomerase, as is often done in reviews on protein classification, because of the simplicity and beauty of its structure (Branden 1980; Richardson 1981; Rossmann and Argos 1981). Its folding pattern is referred to as the "TIM barrel", a compact hydrophobic

Table 1. Domain size in proteins of a single domain

Protein	Number of residues	Reference
Myoglobin	153	Kendrew et al. (1960)
Lysozyme	129	Blake et al. (1965)
Ribonuclease	124	Kartha et al. (1967)
Carboxypeptidase A	307	Lipscomb et al. (1970)
Carbonic anhydrase	258	Liljas et al. (1972)
Adenylate kinase	194	Schulz et al. (1974)
Soybean trypsin inhibitor	181	Sweet et al. (1974)
Triose phosphate isomerase	247	Banner et al. (1975)
Concanavalin A	237	Reeke et al. (1975)
Dihydrofolate reductase	162	Matthews et al. (1978)
Glutathione peroxidase	178	Ladenstein et al. (1979)
Ferritin	174	Heustersprente and Crichton (1981)
Aldolase	225	Mavridis et al. (1982)
Superoxide dismutase	151	Tainer et al. (1982)
Retinol binding protein	182	Newcomer et al. (1984)
α-Crystallin	174	Tardieu et al. (1986)

core of eight parallel β-strands surrounded by eight parallel α-helices. However, in this review, it heads the list for an additional reason. This molecule is a sphere with a diameter of 3.5 nm (Banner et al. 1975), and exemplifies a fundamental concept: the correspondence in size of a protein domain and a water cluster. It is thus the domain volumes, not the chain lengths, that I would prefer to list, but this information is not available.

Kendrew et al. (1958) give dimensions for the myoglobin molecule occupying a space of about 36 nm^3. On the other hand, Perutz et al. (1960) give dimensions for hemoglobin indicating about 44 nm^3 per subunit chain. Since structurally speaking these chains are virtually identical, we conclude that the general myoglobin domain occupies about 40 nm^3. From the papers cited in Table 1 we have the following data: along with triose phosphate isomerase, soybean trypsin inhibitor is 3.5 nm in diameter; crystallin is 3.7 nm; glutathione peroxidase is 3.8 nm; and retinol binding protein is 4.0 nm. The dimensions of aldolase and superoxide dismutase give volumes of just 40 nm^3 each, although these two chains differ in length by 74 residues. Similarly, the shapes of both concanavalin A and carboxypeptidase A, although 70 residues different, have outer dimensions of 4 × 4 × 4 nm. So, despite the wide range of sizes judged from chain length, all these single-domain proteins are folded in such a way that they occupy roughly the same volume in the crystals, about 40 nm^3. Exceptions would appear to be lysozyme and ribonuclease; however, these small molecules must occupy volumes larger than expected from their chain lengths, because both have very open structures with wide clefts to accommodate their large macromolecular substrates, polysaccharides and polynucleotides, respectively.

Carboxypeptidase A is the largest single-domain protein so far reported, and one wonders whether it represents the maximum possible size. Matthews et al. (1972) suggested the size of about 16 000 Da (150 residues) is a convenient one for the optimum polar surface area to nonpolar core volume ratio. Edelman (1973) proposed the "domain hypothesis" on the basis of the genetic control of antibody expression. Wetlaufer (1973) took nucleation of folding events as the underlying reason for domains requiring a size of 40 up to 150 residues, but this was criticized later by Reeke et al. (1975), who pointed out that concanavalin A is a single domain of 237 residues, and so these latter workers took a step further to ask if an upper limit exists at all. Lipscombe et al. (1970) give a molecular mass of 34 600 Da for the large carboxypepsidase A, and using a value of 0.74 for protein specific volume (Matthews 1977) gives a volume of 43 nm^3, corresponding to that of a water cluster.

The majority of proteins whose structures are known are composed of two or more separate domains, many of which are smaller than those listed in Table 1. The proteolytic enzymes (the first group in Table 2) are the most studies and discussed X-ray structures. Each is a monomeric molecule which is folded into two separate halves forming a duplex of two small domains. The middle group in Table 2 lists pairs of separate chains. In their functional state these chains are folded into small domains of about 100 residues each, which associate noncovalently to form dimers. Of course, association of native proteins is not restricted to these small domains. Association is a common and basic form of protein behavior, which does not depend on monomer size or type; for instance, the crystallins and ferritins are totally unrelated in their folded structure but form large multimer aggregates for their function. The small domains exemplified in Table 2 need to be associated in pairs to form an active molecule. They are thus named "semi-domains" to indicate that they are only half of a functional entity. The term "half-domain" is already in use in the

Table 2. Semi-domain size in proteins of one or two domains

Protein	Number of residues	Reference
Papain	111 + 101	Drenth et al. (1970)
Elastase	125 + 120	Shotton and Watson (1970)
Chymotrypsin	123 + 122	Birktoft and Blow (1972)
Thermolysin	156 + 160	Colman et al. (1972)
Trypsin	121 + 124	Stroud et al. (1974)
Penicillopepsin	125 + 117	James and Sielecki (1983)
Fab variable domain light, heavy chain	~110, ~110	Edelman (1973)
Fab constant domain light, heavy chain	~110, ~110	Edelman (1973)
Barnase, barstar	110, 89	Hartley and Smeaton (1973)
Prealbumin	127, 127	Blake et al. (1978)
Nerve growth factor	113, 113	Thoenen and Barde (1980)
Glycoprotein hormone α, β subunit	93, 113	Pierce and Parsons (1981)
Cytochrome C	104	Dickerson et al. (1971)
Thioredoxin	108	Eklund et al. (1984)
Calmodulin	148	Babu et al. (1985)

In the first 6 entries the "+" indicates that the protein is a single chain and so reads N-terminal + C-terminal residues. Edelman gives light and heavy chain fragments of Fab portions of immunoglobulins ranging from 102 to 114.

literature to describe a different concept, as referred to below. The last three entries, cytochrome C, thioredoxin, and calmodulin, belong to widely differing protein families. Although they are usually considered to be monomeric, they are invariably associated with another protein in their active forms.

Semi-domain subunits may be identical, as with prealbumin and nerve growth factor, although heterogeneous association is more common. Although their three-dimensional structures are not known, the subunits of the large family of glycoprotein hormones are tightly associated as dimers. In contrast, the even larger family of growth hormones appear to be active as monomers. However, these are larger molecules, being 191 residues long (Li et al. 1973), and therefore fall into the category of single whole domains.

Examples of possible arrangements of domains, some quite complicated, into which a single chain can fold, are given in Table 3. The first entry is a bifunctional enzyme composed of two similar domains which catalyze two successive steps in a metabolic reaction sequence. Alcohol dehydrogenase and actin are each composed of two dissimilar domains. However, the first belongs to an enormous superfamily which includes kinases, phosphorylases and isomerases, while the ubiquitous actin appears to be a unique molecular species. The serum albumins are composed of three very similar domains, while hydroxybenzoate hydroxylase is folded into one domain followed by two semi-domains. At present, little is available on the three-dimensional structures of molecules larger than 70 000 Da, although one example is the widely distributed transferrin, which is made of four roughly equal domains.

Many domains can be further subdivided on the basis of recurring elements of secondary structure, e.g., the semi-domains of the proteolytic enzymes can be further subdivided into half-domains (McLachlan 1979). These recurring folding units are a much-discussed topic (Matthews 1977; Andreeva and Gustchina 1979; Branden 1980; Ptitsyn and Finkelstein 1980; Rossman and Argos 1981; Richardson 1981; Janin and Wodak 1983; Chothia 1984). It is agreed that the repetitions are the outcome of gene duplications, which increased the sizes of earlier versions of functional proteins and so underlie the general process of their evolution. This theory implies a fundamental point: namely, that proteins are the size they are because of

Table 3. Domain sizes in multidomain proteins

Protein	Number of residues	Reference
Phosphoribose-anthranilate isomerase-indol-glycerol-phosphate synthase	255 + 197	Priestle et al. (1987)
Liver alcohol dehydrogenase	165 + 209	Branden et al. (1973)
Actin	150 + 255	Kabsch et al. (1985)
Human serum albumin	192 + 193 + 200	McLachlan and Walker (1977)
p-Hydroxybenzoate hydroxylase	180 + 108 + 103	Wierenga et al. (1979)
Transferrin	175 + 175 + 175 + 175	Baker et al. (1987)

random events at the level of the gene. Clearly, gene duplication is an efficient mechanism for producing a protein sequence containing elements of repeating structure; but this does not make it the cause. As far as I know, there is no evidence that a quarter of the trypsin molecule binds lysine, or that a quarter of the calmodulin molecule binds Ca ions, or that the βαβ-fold or supersecondary structure binds ATP. I think rather that the whole of each of these molecules was always required for their function; and further, that protein size is therefore determined by other forces which constitute the underlying cause of order in the biological world.

Clegg (1979) has pointed out that the absence of macromolecules smaller than 10 000 Da in the cell poses the question of the significance of size in molecular function. In the cluster-domain model, the shapes and sizes of globular proteins are spatially compatible with water clusters (Watterson 1987c, 1988a). Proteins possess dimensions which allow them to pack mutually together and build the large-scale integrated assembly of protein and solvent we know as the cytoplasmic gel. The gel is not infinitely cross-linked. On the contrary, the separate building blocks, water clusters, and protein domains fit together as replaceable parts. Thus the assembly is flexible, but it is also fragile. Water clusters are noncovalently linked internally and so are readily disrupted. As a consequence, the cytoplasmic superstructure can be destroyed by the mildest of ions or detergents, much less cell homogenization.

In the neighborhood of the cell membrane, water clusters are also ordered in the 3-nm-thick hydration layer. The lipid bilayer is itself 3 nm thick and this dimension no doubt gives stability to the mutual packing of layers of lipid and solvent, just as with the protein assemblies. In addition, many membrane-bound proteins have globular portions with enzyme function. Although anchored in the membrane, these enzymic portions are located 3 nm away from the membrane surface via a connecting stalk or rod (MacLennan et al. 1985; Semenza 1986). This mode of construction places the globular domains beyond the hydration layer and so is further demonstration that the dimensions of hydration fit those of protein domains. These observations lead us to a picture of the membrane, not as a single lipid bilayer, but as orderly stacked layers of lipid bilayers, water clusters, and protein domains.

Further out in the extracellular environment, away from the influence of cellular structures, there is no ordering of clusters because in bulk water the chaos of disruptive thermal fluctuations prevails. Proteins bound for this environment are strengthened in their folded conformations by internal covalent disulfide crossbridges. Thus the lack of the stable integrated superstructure is the reason for the presence of these internal crossbridges in certain proteins, not the thermodynamics of special folding mechanisms.

3.2 Domain Energetics

The subcellular medium is mechanically very weak, yet at the same time it is highly energized. These strongly contrasting physical properties are indeed thought-provoking since, without a strong containing framework, they are a recipe for chaotic disaster. In the cluster-domain model, the structural stability of the medium is due to the packing of spatially compatible units. This is a co-operative process, so that the larger the superstructure, the more stable are its components, thus ensuring the stability of folded protein chains. This idea contrasts with the reductionist view of statistical thermodynamics, because it attributes protein stability to forces operating from above on the large scale, and not to the summation of independent contributions from separate small molecular effects (Watterson 1988b).

The binding of S-peptide, the first 20 N-terminal residues of ribonuclease A, to the rest of the protein illustrates this point. S-peptide is produced by cleaving the intact chain with subtilisin, but binds to the remainder of the molecule with high affinity under native conditions. This complex is active, while the truncated enzyme without S-peptide is inactive (Richards and Vithayathil 1959). The three-dimensional structure of the complex (Wyckoff et al. 1967) reveals only slight and superficial contact, whereby the reside Asp 14 at about the middle of the peptide is described as making a charge-charge interaction, or ionic bond, with the rest of the molecule. However, an ionic bond does not possess the strength of a covalent linkage, and so the complex could not survive energetic fluctuations such as expected during activity. In the cluster model, the binding forces originate from outside the domain. They travel through the water, into and through the protein molecule, so that all sections of the molecule are held in place by a force imposed on them by the large-scale structure of which they are part, and not by small-scale forces which are part of them.

That the cluster size is a fundamental energetic unit is further illustrated by the tight association of semi-domains. Small proteins dimerize often without an indication of what causes this ordering process. In their comparison of immunoglobulin-variable domains, Novotny and Haber (1985) examine the interaction across the interface between heavy and light chain semi-domains. One expects perhaps here, more than in any other case of protein association, to find obvious binding forces because of the finely tuned specificity shown by antibody function. Yet these workers found hydrogen bonding between just one residue on the light chain (Gln 38) and one on the heavy chain (Gln 39) to "extend the hydrogen bonded network across the domain-domain interface and anchor the interface β-sheets in their relative orientation." When one considers that there must be at least 100 hydrogen bonds within each separate chain of 100 residues (Baker and Hubbard 1984) which hold its conformation in place, then one must attribute very special binding powers indeed to this single interaction between the chains, if such small-scale polar interactions are seen as the only agents that can produce force. On the other hand, the cluster model provides a simpler explanation. Isolated molecules of the semi-domain size do not pack well geometrically into the solvent cluster network and consequently it is energetically favorable for them to double their size.

That the force underlying protein-protein attraction operates on a scale as large as the domain is a direct consequence of the cluster-domain model. The co-operative making and breaking of hydrogen bonds, that transmits the structure wave through water, operates equally well throughout protein domains. In Fig. 6, the domain, depicted as a barrel of twisted β-strands, is interpreted as a collection of hydrogen bonds holding the strands together. These rows of flexible bonds open and

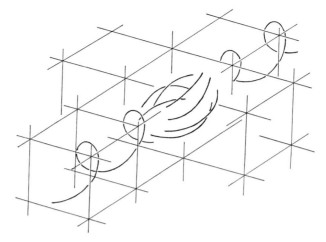

Fig. 6. Diagrammatic illustration of an array of clusters each occupying a volume of approximately 40 nm^3. The *central cube* contains a group of twisted strands representing a globular protein domain. The *spiral* represents the structure wave passing back and forth without interruption through the clusters and the domain

close in zipper-like fashion as the wave passes through the structure (Watterson 1988b). When two neighboring domains, whether belonging to the same molecule or not, adopt a mutual orientation so that the wave can cross their interface, the wavelength can double in the same way as depicted for clusters in Fig. 5. As a result, the internal tension that holds a domain together now operates through the network of bonds in both at once. As with the clusters, these harmonic transitions can produce multidomains of extensive proportions, able to transmit tension on a large scale. In this model, the apparent duplication seen in protein folding anatomy, as in half-domains, semi-domains, whole domains and so on, is intimately connected with the periodic nature of wave motion.

We can compare both approaches, that based on classical statistical theory and that of the cluster-domain model, by taking some well studied examples of protein interactions. Two very different systems, the soybean trypsin inhibitor-trypsin complex and the lysozyme-antilysozyme complex, are considered to be well understood on the basis of their crystal structures. In the first case, Sweet et al. (1974) give the dimensions of the inhibitor molecule as a sphere of 3.5 nm diameter and so in shape and size it resembles trypsin itself. The two spheres share a region of little contact in comparison to their overall sizes, and the authors single out 5 or 6 residues, out of a total of 400 or more, which are responsible for the high affinity interaction. The picture is a similar one in the case of the antigen-antibody complex. Amit et al. (1987) find that 16 residues of lysozyme make contact with 17 of its antibody across a "rather flat surface". However, these contacts are not described as bonds, and the authors list separately 12 hydrogen bonds which must hold together a sphere of 15 000 Da onto the end of an ellipsoid of 50 000 Da in a precise manner.

In the cluster approach, the solvent is an active participant in protein interaction. The attractive force holding the complex together could not operate without the presence of the surrounding medium. The force is transmitted through the whole of the space defined by the dimensions of the fused domains, whether the area of actual protein-protein contact is large or small. Thus, the explanation of the high affinity interactions lies in this large-scale force, and not in a few noncovalent bonds of extraordinary strength.

The dynamic properties mean that the domain-domain or domain-cluster interface has the potential of being a very active region. This follows because when the step doubling the wavelength across two domains occurs, the interface now experiences forces where previously there were none. For the transition to take place, the opposed surfaces need to be compatible, in the sense that the networks of hydrogen bonding existing within each domain separately can be joined, so that overall co-operativity remains ensured and the structure wave can pass smoothly across the interface. Adjacent surfaces can be made compatible by insertion of a small molecule, e.g., substrate, cofactor or metal ion, which makes bonds to both sides, thus bridging the interface. The transition can now take place, and as a result this small molecule will experience a tensile force pulling on it via these interconnecting bonds. It is the same force that pulls water across the membrane in osmosis. According to the amplitude of the ideal stationary wave depicted in Fig. 7, the tension becomes maximum at the center of the fused domain. It is now an easy step to propose that the large oscillations in tension precisely in this region can cause the making and breaking of covalent bonds within the small molecule. In other words, the cluster-domain model provides a mechanism in which chemical events at the enzymic active site are physically coupled to the mechanical events of larger scale in the surroundings. Furthermore, this coupling need not be localized to the region of one molecule. We have seen how clusters and domains can be co-operatively linked by transitions in the structure wave. The linkage of series of enzymic sites in an integrated super-structure opens the way for ordered interplay between the metabolism of the cell and the changing physical states of the cytoplasm.

At present, protein structure, metabolic sequences, and cytoplasmic streaming are regarded as disparate subjects, with the fact that they are all biological being their single common feature. It is, of course, possible that these different features of living systems are unconnected, that each operates independently according to its own principles, and that the cell functions by adopting average situations which result from summation of all the independent processes operating at a given moment. It is, however, unlikely that this mode of function could explain the ordered move-

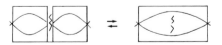

Fig. 7. The diagram represents in an extremely simplified way the functioning of an enzymic active site. When the substrate molecule, represented by the *zig-zag line*, is inserted between two domains, the harmonic transition becomes possible and the oscillating forces now acting in this region split it to form the products

ment exhibited by cells. Subcellular movement takes place as though directed by an underlying co-ordination, implying a unifying principle which links metabolic chemical energy reciprocally with macroscopic mechanical forces. This principle is clearly one of structure existing throughout subcellular space, and of all subcellular components, I think that water is the only one capable of fulfilling this role.

References

Amit AG, Mariuzza RA, Phillips SEV, Puljak RJ (1987) Three-dimensional structure of an antigen-antibody complex at 2.8A° resolution. Science 233:747–753

Amos LA, Amos WB (1987) Cytoplasmic transport in axons. J Cell Sci 87:1–2

Andreeva NS, Gustchina AE (1979) On the supersecondary structure of acid proteases. Biochem Biophys Res Commun 87:32–42

Andrews FC (1976) Colligative properties of simple solutions. Science 194:567–571

Babu YS, Sack JS, Greenhough TJ, Bugg CE, Means AR, Cook WJ (1985) Three-dimensional structure of calmodulin. Nature (Lond) 315:37–40

Baker EN, Hubbard RE (1984) Hydrogen bonding in globular proteins. Prog Biophys Mol Biol 44:97–179

Baker EN, Rumball SV, Anderson BF (1987) Transferrins. Trends Biochem Sci 12:350–353

Banner DW, Bloomer AC, Petsko GA, Phillips DC, Pogson CI, Wilson IA, Corran PH, Furth AJ, Milman JD, Offord RE, Priddle JD, Waley SG (1975) Structure of chicken muscle triose phosphate isomerase determined crystallographically at 2.5 A° resolution using amino acid sequence data. Nature (Lond) 255:609–703

Birktoft JJ, Blow DM (1972) Structure of crystalline α-chymotrypsin. J Mol Bio 68:187–240

Blake CCF, Koenig DF, Mair GA, North ACT, Phillips DC, Sarma VR (1965) Structure of hen egg white lysozyme. Nature (Lond) 206:757–761

Blake CCF, Geisow MJ, Oatley SJ, Rerat B, Rerat C (1978) Structure of prealbumin. J Mol Biol 121:339–356

Branden C (1980) Relation between structure and function of α/β proteins. Quart Rev Biophys 13:317–338

Branden C, Eklund H, Nordstrom B, Boiwe T, Soderlund G, Zeppezauer E, Ohlsson I, Akeson A (1973) Structure of liver alcohol dehydrogenase at 2.9 A° resolution. Proc Natl Acad Sci USA 70:2439–2442

Buxbaum RE, Dennerll T, Weiss S, Heidemann SR (1987) F-actin and microtubule suspensions as indeterminate fluids. Science 235:1511–1514

Chothia C (1984) Principles that determine the structures of proteins. Ann Rev Biochem 53:537–572

Churaev NV, Derjaguin BV (1985) Inclusion of structural forces in the theory of stability of colloids and films. J Coll Interface Sci 103:542–553

Clegg JS (1979) Metabolism and the intracellular environment, the vicinal water network model. In: Drost-Hansen W, Clegg JS (eds) Cell-associated water. Academic Press, NY, pp 363–380

Clegg JS (1984) Properties and metabolism of the aqueous cytoplasm and its boundaries. Am J Physiol 246:R133–R151

Colman PM, Jansonius JN, Matthews BW (1972) The structure of thermolysin. J Mol Biol 70:701–724

Cooper A (1976) Thermodynamic fluctuations in protein molecules. Proc Natl Acad Sci USA 73:2740–2741

Crawford IP, Niermann T, Kirschner K (1987) Prediction of secondary structure by evolutionary comparison. Prot Struct Funct Genet 2:118–129

Dickerson RE, Takano T, Eisenberg D, Kallai DB, Samson L, Cooper A, Margoliash E (1971) Ferricytochrome C. J Biol Chem 246:1511–1521

Drenth J, Jansonius JN, Kockoek R, Sluyterman LAA, Wolthers BG (1970) Cysteine proteinases. Phil Trans Roy Soc Lond B 257:231–236

Drost-Hansen W (1985) Role of vicinal water in cellular evolution. In: Pullman A, Vascilescu V, Packer L (eds) Water and ions in biological systems. Plenum, New York, pp 523–534

Edelman GM (1973) Antibody structure and molecular immunology Science 180:830–840

Eklund H, Cambillau C, Sjoberg BM, Holmgren A, Jornvall H, Hoog JO, Branden C (1984) Conformational and functional similarities between glutaredoxin and thioredoxin. EMBO J 3:1443–1449

Flory PJ (1953) Principles of polymer chemistry. Cornell Univ Press, Ithaca, pp 348–398

Franks F (1975) The hydrophobic interaction. In: Franks F (ed) Water. A comprehensive treatise, vol 4. Plenum Press, New York, pp 1–64

Frauenfelder H (1983) Summary and outlook. In: Porter R, O'Conner M, Wehlan J (eds) Mobility and function in proteins and nucleic acids. Pitman, London pp 329–339 (Ciba Foundation Symposium, vol 93).

Frauenfelder H, Parak F, Young RD (1988) Conformational substates in proteins. Ann Rev Biophys Biophys Chem 17:451–479

Gaeta FS, Mita DG (1979) Thermal diffusion across porous partitions. J Phys Chem 83:2276–2285

Gershon ND, Porter KR, Trus BL (1985) The cytoplasmic matrix. Proc Natl Acad Sci USA 82:5030–5034

Giordano R, Salvato G, Wanderlingh F, Wanderlingh U (1990) Quasielastic and inelastic neutron scattering in macromolecular solutions. Phys Rev A 41:689–696

Glasstone S, Lewis D (1963) Elements of physical chemistry. Macmillan, London pp 248–251

Goldenberg DP (1988) Genetic studies of protein stability and mechanisms of folding. Ann Rev Biophys Biophys Chem 17:481–507

Hammel HT, Scholander PF (1976) Osmosis and tensile solvent. Springer, Berlin Heidelberg New York

Hartley RW, Smeaton JR (1973) On the reaction between the extracellular ribonuclease of *Bacillus amyloliquefaciens*, barnase, and its inhibitor, barstar. J Biol Chem 248:5624–5626

Heustersprente M, Crichton RR (1981) Amino acid sequence of horse spleen apoferritin. FEBS Lett 129:322–330

Hildebrand JH (1968) A criticism of the term "hydrophobic bond." J Phys Chem 72:1841–1842

Hildebrand JH (1979) Forum on osmosis. A criticism of solvent tension in osmosis. Am J Physiol 237:R108–R109

Huxley HE (1973) Muscle contraction and cell motility. Nature (Lond) 243:445–449

Israelachvilli JN, Adams GE (1976) Direct measurements of long range forces between two mica surfaces in aqueous potassium nitrate solutions. Nature (Lond) 262:774–776

James MN, Sielecki AR (1983) Strucure and refinement of penicillopepsin at 1.8A° resolution. J Mol Biol 163:299–361

Janin J, Wodak SJ (1983) Structural domains in proteins and their role in the dynamics of protein function. Prog Biophys Mol Biol 42:21–78

Kabsch W, Mannherz HG, Suck D (1985) Three-dimensional structure of the complex of actin and DNase I at 4.5 A° resolution. EMBO J 4:2113–2118

Kaprelyants AS (1988) Dynamic spatial distribution of proteins in the cell. Trends Biochem Sci 13:43–46

Karplus M, McCammon JA (1983) Dynamics of proteins, elements and function. Ann Rev Biochem 52:263–300

Kartha G, Bello J, Harker D (1967) Tertiary structure of ribonuclease. Nature (Lond) 213:862–865

Kauzmann W (1959) Some factors in the interpretation of protein denaturation. Adv Prot Chem 14:1–63

Kell DB (1988) Coherent properties of energy coupling membrane systems. In: Fröhlich H (ed) Biological coherence and response to external stimuli. Springer, Berlin Heidelberg New York Tokyo, pp 233–241

Kempner ES, Miller JH (1968) The molecular biology of *Euglena gracilis* IV. Cellular stratification by centrifugation. Exp Cell Res 51:141–149

Kendrew JC, Bodo G, Dintizis HM, Parrish RG, Wyckoff H, Phillips DC (1958) A three-dimensional model of the myoglobin molecule obtained by X-ray analysis. Nature (Lond) 181:662–666

Kendrew JC, Dickerson RE, Standberg BE, Hart RG, Davies DR, Phillips DC, Shore VC (1960) Structure of myoglobin. Nature (Lond) 185:422–427

Ladenstein R, Epp O, Bartels K, Jones A, Huber R, Wendel A (1979) Structure analysis and molecular model of the selenoenzyme glutathione peroxidase at 2.8 A° resolution. J Mol Biol 134:199–218

Lamanna R, Cannistraro S (1989) Water proton self-diffusion and hydrogen bonding in aqueous human albumin solutions. Chem Phys Lett 164:653–656

LeNeveu DM, Rand RP, Parsegian VA (1976) Measurement of forces between lecithin bilayers. Nature (Lond) 259:601–603

Li CH, Gordon D, Knorr J (1973) The primary structure of sheep pituitary growth hormone. Arch Biochem Biophys 156:493–508

Liljas A, Kannan KK, Bergsten PC, Waara I, Fridborg K, Stanberg B, Carlbom U, Jarup L, Lovegren S, Petel M (1972) Crystal structure of huamn carbonic anhydrase. Nature New Biol 235:131–137

Lipscomb WN, Reeke GN, Hartsuck JA, Ouiocho FA, Bethge PH (1970) The structure of carboxypeptidase A. Phil Trans Roy Soc Lond B 257:177–214

Luby-Phelps K, Lenni F, Taylor DL (1988) The submicroscopic properties of cytoplasm as a determinant of cellular function. Ann Rev Biophys Biophys Chem 17:369–396

MacLennan DH, Brandl CJ, Korczak B, Green NM (1985) Amino acid sequence of Ca + Mg-dependent ATPase from rabbit muscle sarcoplasmic reticulum deduced from its complementary DNA sequence. Nature (Lond) 316:696–700

Masters CJ, Reid S, Don M (1987) Glycolysis. A new concept in an old pathway. Mol Cell Biochem 76:3–14

Matthews BW (1977) X-ray structure of proteins. In: Neurath H, Hill R, Boeder C (eds) The proteins, vol 3. Academic Press, New York, pp 403–590

Matthews BW, Jansonius JN, Cloman PM, Schuenborn BP, Dupourque D (1972) Three-dimensional structure of thermolysin. Nature New Biol 238:37–41

Matthews BW, Alden RA, Bolin JT, Filman DJ, Freer ST, Hamlin R, Hol WGJ, Kisliuk RL, Pastore EJ, Plante LT, Xuong N, Kraut J (1978) Dihydrofolate reductase from *Lactobacillus casei*. J Biol Chem 253:6946–6954

Mavridis IM, Hatada MH, Tulinsky A, Lebioda L (1982) Structure of 2-keto-3-deoxy-6-phosphogluconate aldolase at 2.8 A° resolution. J Mol Biol 162:419–444

McLachlan AD (1979) Gene duplication in the structural evolution of chymotrypsin. J Mol Biol 128:49–79

McLachlan AD, Walker JE (1977) Evolution of serum albumin. J Mol Biol 112:543–558

Moore WJ (1956) Physical chemistry, 2nd edn. Longmans, London, pp 48–51

Newcomer ME, Jones TA, Aquist J, Sundelin J, Eriksson U, Rask L, Peterson PA (1984) The three-dimensional structure of retinol binding protein. EMBO J 3:1451–1454

Ninham BW (1980) Long-range versus short-range forces. J Phys Chem 84:1423–1430

Norrish K (1954) The swelling of montmorillonite. Disc Faraday Soc 18:120–134

Novotny J, Haber E (1985) Structural invariants of antigen binding. Proc Natl Acad Sci USA 82:4592–4596

Perutz MF (1979) Regulation of oxygen affinity of hemoglobin. Ann Rev Biochem 48:327–386

Perutz MF, Rossmann MG, Cullis AF, Muirhead H, Will G, North ACT (1960) Structure of hemoglobin. Nature (Lond) 185:416–422

Pierce JG, Parsons T (1981) Glycoproptein hormones, structure and function. Ann Rev Biochem 50:465–495

Plumb RC, Bridgeman WB (1972) Ascent of sap in trees. Science 176:1129–1131

Porter KR, Berkerle M, McNiven A (1983) The cytoplasmic matrix. Mod Cell Biol 2:259–273

Priestle JD, Grutter MG, White JL, Vincent MG, Kania M, Wilson E, Jardetzky TS, Kirschner K, Jansonuis JN (1987) Three-dimensional structure of the bifunctional enzyme N-5-phosphoribosyl anthranilate isomerase — indole-3-glycerol phosphate synthase from *E. coli*. Proc Natl Acad Sci USA 84:5690–5694

Privalov PL (1979) Stability of proteins. Adv Prot Chem 33:167–241

Ptitsyn OB, Finkelstein AV (1980) Similarities of protein topologies. Ouart Rev Biophys 13:339–386

Reeke GN, Becker JW, Edelman GM (1975) The covalent and three-dimensional structure of concanavalin A. J Biol Chem 250:1525–1547

Richards FM, Vithayathil BJ (1959) The preparation of subtilisin-modified ribonuclease and the separation of protein and peptide components. J Biol Chem 234:1459–1465

Richardson JS (1981) The anatomy and taxonomy of protein structure. Adv Prot Chem 34:167–339

Rossmann MG, Argos P (1981) Protein folding. Ann Rev Biochem 50:497–532

Scholander PF, Hammel HT, Bardstreet ED, Hemmingsen EA (1965) Sap pressure in vascular plants. Science 148:339–346

Schulz GE, Elizinga M, Marx F, Schirmer RH (1974) Three-dimensional structure of adenylate kinase. Nature (Lond) 250:120–124

Semenza G (1986) Anchoring and biosynthesis of stalked brush border membrane proteins. Ann Rev Cell Biol 2:255–313

Shotton DM, Watson HC (1970) The three-dimensional structure of crystalline porcine pancreatic elastase. Phil Trans Roy Soc Lond B 257:111–118

Srere PA (1987) Complexes of sequential metabolic enzymes. Ann Rev Biochem 56:89–124

Srivastrava DK, Bernhard SA (1987) Biophysical chemistry of reaction sequences. Ann Rev Biophys Biophys Chem 16:175–204

Stroud RM, Kay LM, Dickerson RE (1974) The structure of bovine trypsin. J Mol Biol 83:185–208

Sweet RM, Wright HT, Janim J, Chothia CH, Blow DM (1974) Crystal structure of the complex of porcine trypsin with soybean trypsin inhibitor at 2.6 A° resolution. Biochemistry 13:4212–4224

Tainer JA, Getzoff ED, Beem KM, Richardson JS, Richardson DC (1982) Determination and analysis of the 2 A° structure of copper, zinc superoxide dismutase. J Mol Biol 160:181–217

Tanford C (1968) Protein denaturation. Adv Prot Chem 23:121–282

Tanford C (1973) The hydrophobic interaction. Wiley, New York

Tardieu A, Laporte D, Licinio P, Krop B, Delaye M (1986) Calf lens α-crystallin quaternary structure. J Mol Biol 192:711–724

Taylor DL, Fechheimer M (1982) Cytoplasmic structure and contractility, the solation-contraction coupling hypothesis. Phil Trans Roy Soc Lond B 299:185–197

Thoenen H, Barde YA (1980) Physiology of nerve growth factor. Physiol Rev 60:1284–1335

Van Olphen H (1954) Interlayer forces in bentonite. Clays Clay Miner 2:418–438

Watterson JG (1982) Model for a co-operative structure wave. In: Franks F, Mathias SF (eds) Biophysics of water. Wiley, New York, pp 144–147

Watterson JG (1987a) Does solvent structure underlie osmotic mechanisms? Phys Chem Liq 16:313–316

Watterson JG (1987b) Solvent cluster size and colligative properties. Phys Chem Liq 16:317–320

Watterson JG (1987c) A role for water in cell structure. Biochem J 248:615–617

Watterson JG (1988a) The role of water in cell architecture. Mol Cell Biochem 79:101–105

Watterson JG (1988b) A model linking water and protein structures. Bio Systems 22:51–54

Watterson JG (1989) Wave model of liquid structure in clay hydration. Clays Clay Miner 37:285–286

Weber G (1975) Energetics of ligand binding to proteins. Adv Prot Chem 29:1–83

Welch GR (1977) On the role of organized multienzyme systems in cellular metabolism. Prog Biophys Mol Biol 32:103–191

Welch GR, Somogui B, Damjanovich S (1982) The role of protein fluctuations in enzyme action. Prog Biophys Mol Biol 39:109–146

Wetlaufer DB (1973) Nucleation, rapid folding, and globular intrachain regions in proteins. Proc Natl Acad Sci USA 70:697–700

Wierenga RK, de Jong RJ, Kalk KH, Hol WGJ, Drenth J (1979) Crystal structure of p-hydroxybenzoate hydroxylase. J Mol Biol 131:55–73

Wiggins PM, MacClement BAE (1987) Two states of water found in hydrophobic clefts. Int Rev Cytol 108:249–303

Wyckoff HW, Hardman KD, Allewell NM, Inagami T, Johnson LN, Richards FM (1967) The structure of ribonuclease-S at 3.5 A° resolution. J Biol Chem 242:3984–3988

Subject Index

2-aminopurine 46
Achlya ambisexualis 98
actin filaments 59, 126
actinin 75
adenovirus E1A protein 96
adenylate kinase 124
alcohol dehydrogenase 126
aldolase 124
α-amanitin 3
α-crystallin 124
alternative splicing 19
amphibian nuclei 45
amphibians 66
Arabidopsis 5
ATGCAAAT 15
AUAUAC sequence 22, 24
autophosphorylation 53
AUUUUUG 4

barnase 125

Caenorhabditis 78
calmodulin 125
cAMP response element binding
    protein 15
cap structure and U snRNA 20
carbonic anhydrase 124
carboxypeptidase A 124
Carnot cycle 119
casein 39, 43, 48
casein kinase 1, 46
CCAAAT box 93
cGMP-dependent histone kinase 39
cGMP-dependent kinase 38
Chaetopterus 76
chymotrypsin 125
cluster
    dynamics 122
    energetics 119
    model of liquid structure 116
    size 116
    -domain model 128
concanavalin A 124
contractile ring 66

cortical
    cytoskeletal domain 81
    network 79
creatine phosphokinase 60
cytochalasin B 67
cytoskeletal domain 72
cytoskeleton 59
    and annelids 76
    and ascidians 63
    and early development of chordates 63
    and early development of
        nonchordates 76
    and echinoderms 79
    and nematodes 78
    and oligochaetes 77
    of somatic cells 66

Dictyostelium 1
dihydrofolate reductase 124
distal sequence element 15
DNA polymerase 38
DNA protein kinase
    and autophosphorylation 47
    and cofactor requirements 45
    and comparison with other nuclear
        protein kinases 51
    and effects of polynucleotides 49
    and inhibitors 46
    and phosphate donor 45
    and phosphorylation site 48
    and physical characteristics 43
    and purification 42
    and subcellular localization 45
    and substrate preference 48
    and substrate specificity 47
    and transcription regulation 54
DNA-activated protein kinase 39
DNA-dependent protein kinase 42
DNA-stimulated protein phosphorylation 39
domain
    energetics 127
    model of protein structure 123
    -cluster interface 130
    -domain interface 130

EIA-like cellular factor 95
elastase 125
Escherichia coli 89

Fab variable domain 125
ferritin 124
fodrin 75

gel-sol transition 114
gene promoter 18
  and mRNA 18
  and snRNA 18
germinal vesicle breakdown 38
glutathione peroxidase 124
glyceraldehyde-3-phosphate dehydrogenase 101
glycoprotein hormone 125
grp78 promoter 95

heat-shock factor 95
heat-shock gene 90
  and gametogenesis 90
  and SAP factor 92
heat-shock promoter 101
heat-shock protein 40, 89
  and deficient heat-shock responses 100
  and differentiation 96
  and early embryogenesis 94, 100
  and gametogenesis 100
  and high levels of B2 transcripts 96
  and hormone 97
  hyperexpression 90
  and mouse embryonal carcinoma cells 94
  and sporulation 100
  synthesis 102
  and undifferentiated mouse embryonic cells 96
  and zygote genome transcription 93
helicases 38
heparin 46
Histoplasma capsulatum 91
hsp26 91
hsp70 89, 90, 91, 99, 100
hsp83 91
hsp85 98
hsp87 100
hsp90 transcription 99

initiation nucleotide 13
inorganic phosphate 46
intermediate filament 74
intracellular translocation 59
isomerase-indol-glycerol-phosphate
  synthetase 126

Ku autoantigen 49

La antigen 5
Leishmania 91

liver alcohol dehydrogenase 126
lysozyme 124

m^{2,7}G cap 24
m^{7}A cap 24
m^{7}G cap 24
MepppG 3, 4
MepppG cap structure 20
MepppG/A-capped RNA 21
microtubule 59
  -organizing center 74
mos protein 38
myoglobin 124

N,N-dimethylaminopurine 46
N,N-dimethylaminopurine riboside 46
N-ethylmaleimide 46
nerve growth factor 125
NI kinase 37
NII kinase 37
Nitella 60
nuclear protein kinase 37
nucleases 38

octamer-binding element 16
osmosis 116

P granules 78
p-hydroxybenzoate 126
papain 125
penicillopepsin 125
permeable membrane 116
phosphoribose-anthranilate 126
phosphorylation
  cascade 37
  event 53
phosvitin 45
polyadenylation 18
polymeric solute 122
pp60^{src} 38
protein and cellular function 113
proximal sequence element 13
pyrophosphate 46

quercetin 46

reticulocyte lysate 45
retinoic acid 98
retinol binding protein 124
ribonuclease 124
RNA polymerase 3, 20, 38
rutin 46

7SK RNA 2, 23
Saccharomyces 90
serum albumin 126
signal transduction 37

Sm antigen 4
Sm-protein 24
small nuclear RNA 1
snRNA 1
  regulation 19
  synthesis 19
solvent clusters 118
solvent-solvent interaction 117
soybean trypsin inhibitor 124
Spl phosphorylation 49
spectrin 75
Styela 63
superoxide dismutase 124
SV40 l-antigen 38

tagetitoxin 5
TATA box 14
TFIIIA 16
TFIIIB 15, 16
TFIIIC 16
thermal fluctuations in proteins 115
thermolysin 125
thioredoxin 125
threonine 44
TMG-capped 4
  -snRNA 17
TMGpppA/-capped, RNA 21
trans-acting factors 15
transcription factor TFIIIB 96
trimethylguanosine (TMG) cap 1
triose phosphate isomerase 124
Trypanosoma 91
trypsin 125

U snRNA 2, 3
  promoter 7

and trans-acting factors 15
U snRNA genes 5
  and chicken 10
  and *Drosophila* 11
  and human 5
  and plants 12
  and rodents 10
  and sea urchin 11
  and *Trypanosome* 11
  and viral U RNAs 12
  and *Xenopus* 10
  and yeast 12
U snRNP protein 24
U1 RNA 2, 4
U2 RNA 2, 4
U3 RNA 2
U4 RNA 2
U5 RNA 2
U6 RNA 2
  and post-transcriptional
    capping 22
U7 RNA 2
U8 RNA 2
U11 RNA 2
U12 RNA 2
U13 RNA 2
U14 RNA 2
ubiquitin 91, 100

vinculin 75

water 113

yolk-synthesis stage 68
YYCAYYYY 3' 13

Printing: COLOR-DRUCK DORFI GmbH, Berlin
Binding: Buchbinderei Lüderitz & Bauer, Berlin